ADVANCED TECHNOLOGIES AND APPLICATIONS OF NEUTRON ACTIVATION ANALYSIS

Edited by **Lylia Alghem Hamidatou**

Advanced Technologies and Applications of Neutron Activation Analysis

http://dx.doi.org/10.5772/intechopen.69725
Edited by Lylia Alghem Hamidatou

Contributors

Lylia Alghem Hamidatou, Guillermo Abel Parrado, Mary Peña-Urueña, David Alonso-Contreras, Mauricio López-Rodríguez, Andrés Porras-Rios, Martha Guzmán-Guzmán, Julián Orozco-García, Nelson Acero-Barriga, Maitreyee Nandy, Vitaly Krasilnikov, MunSeong Cheon, Luciano Bertalot, Hyeon Gon Lee, Pham Ngoc Son, Bach Nhu Nguyen, Maria Angela De Barros Correia Menezes, Radojko Jacimovic

Notice

First published in London, United Kingdom, 2019 by IntechOpen
IntechOpen is the global imprint of INTECHOPEN LIMITED, registered in England and Wales, registration number: 11086078, The Shard, 25th floor, 32 London Bridge Street
London, SE19SG – United Kingdom
Printed in Croatia

British Library Cataloguing-in-Publication Data
A catalogue record for this book is available from the British Library

Additional hard copies can be obtained from orders@intechopen.com

Advanced Technologies and Applications of Neutron Activation Analysis, Edited by Lylia Alghem Hamidatou
p. cm.
Print ISBN 978-1-78985-973-7
Online ISBN 978-1-78985-974-4

For EU product safety concerns:
IN TECH d.o.o., Prolaz Marije Krucifikse Kozulić 3, 51000 Rijeka, Croatia,
info@intechopen.com or visit our website at intechopen.com.

Meet the editor

Dr. Lylia Alghem Hamidatou has been the director of the Nuclear Techniques and Applications Division and a senior research scientist at the Nuclear Research Centre of Birine CRNB, Algerian Atomic Energy Atomic Commission COMENA, since 1997.

In September 2017, she was elected as a new member of the k0-International Scientific Committee. She was granted "Habilitation" accreditation to direct research in April 2014. In 2010, she was awarded with a distinction for her doctoral degree in Nuclear Physics at the University of Ferhat Abbes of Sétif-Algeria. Her research interests include neutron activation analysis (NAA), development and application of several techniques such as k0-NAA and radiochemical NAA, and cyclic-delayed neutron counting. In 2008, she won the Young Scientist Award in recognition of her early career achievements in nuclear analytical chemistry by the International Committee on Nuclear Analytical Methods in the Life Sciences at NAMLS-9 in Lisbon, Portugal. She has published more than 45 articles in various international journals, books, and conference proceedings. Her research interests are focused on the development of the PGAA method at the Es-Salam research reactor, the study of several types of matrices using k0-NAA and relative methods (major and trace elements determination in Algerian Alzheimer's, psoriatic, and breast cancer patients), toxic and essential elements in seeds and medicinal plants, and rare earth elements in geological samples (rocks, clays, marble, sediments, etc.). She has participated in many IAEA_RAF regional projects to assess the analytical performance of the Algerian Neutron Activation Analysis Laboratory at CRNB.

Contents

Preface

Since the last century, physicists have known that most of the mass of an atom was located in the nucleus at its center and that this central core contained protons. Until 1932, the atom was believed to be composed of a positively charged nucleus surrounded by negatively charged electrons. Atomic theory from James Chadwick's discovery of the neutron was central to the extraordinary developments in atomic physics that occurred in the first half of the twentieth century. The neutron was exploited to create new radioactive elements by neutron irradiation and the fission of uranium atoms.

Hevesy and H. Levis were the first to report on a new method of activation analysis. Today, the concept of neutron activation analysis (NAA) has developed into a research reactor characterized with high neutron fluxes and gamma ray spectrometers with high resolving power HP(Ge) and NaI(Tl) detectors coupled to full-featured multichannel analyzer-based advanced digital signal processing techniques. The complete spectroscopy workstation offers the highest quality acquisition and analysis. The use of NAA by researchers has increased steadily, and as a result, this technique remains one of the most common activities in 240 existing nuclear research reactors worldwide. In October 2017, the IAEA launched the first e-learning course dedicated to a research reactor-based technique and made it available to Member States through the IAEA Open Learning Management System.

This book highlights the advanced technologies and applications of NAA. It discusses the latest developments influencing the performance and utility of different NAA techniques across wide areas of applications: nuclear technology, industry, medicine, etc. with accurate analytical results. The overall goal of the book is to promote innovation and development of NAA techniques, technologies, and nuclear culture by presenting high-quality chapters with numerous results at both national and international levels. Since the k_0-NAA method was launched by A. Simonits and F. De Corte, many laboratories worldwide have implemented and applied this technique.

The book is intended to provide guidance in practical aspects of operating and applying laboratory NAA and reviews of the recent applications and associated technologies. It provides further insight into the method of analyzing samples using INAA, k_0-NAA, and LS-NAA, which are sensitive analytical techniques useful for performing both qualitative and quantitative analyses.

It is hoped that the book will serve as a source for graduate and postgraduate students in nuclear sciences and applications and nuclear analytical techniques, as well as teachers and researchers who have an overlapping interest in recent and projected needs of research and development. The neutron activation system was designed for ITER to evaluate integrated fusion power and develop an integrated approach to routine automation of NAA, and has resulted in interesting applications from several fields of science: medicine, clinical investigations, biology, geochemistry, soil contamination, waste management, diet, lifestyle and health, cosmology, archeology, forensic science, etc.

The chapters cover the basis of the NAA method, which includes principles and properties, the use of a Monte Carlo computational model for many case studies, scientific achievements, and potential applications of NAA performed at the author's laboratory and at nuclear facilities. Readers should refer to the work discussed in all chapters.

The book is organized into two sections, which cover a variety of very important topics. The first section contains four chapters on different applications of NAA in several fields of research. The first chapter by Lylia Hamidatou presents an overview of NAA methods, including innovative worldwide applications. In the second chapter, Maitreyee Nandy highlights the application of neutron activation and subsequent gamma spectrometric studies in the fields of geology and medicine.

Different types of samples such as uranium ores, phosphate rocks, cement, beach rocks, ancient lake Copais samples, and many other kinds of geological matrix were used for the determination of major, trace, ultra-trace, and rare earth elements using neutron activation followed by high-resolution gamma spectrometric studies. The author also discusses the different applications of NAA in medicine. Concentrations of trace levels of Ca, Cu, Co, I, Mg, Se, Fe, Zn, Hg, Ba, and Cr in human breast cancer, skin cancer, and colorectal cancer tissues and in the dysfunction and malignancy of the thyroid gland were determined using the NAA method.

Chapter 3 by Guillermo Parrado, David Alonso, Julián Orozco, Mary Peña, Andrés Porras, Martha Guzmán, Nelson Acero, and Mauricio López reports on the effort of a Colombian NAA laboratory and potential applications and developments using the nuclear research reactor IAN-R1. The Colombian laboratory serves as a good example, not only for novice labs, but also for the more experienced ones. Lessons have been learned by exploiting these young scientists and engaging them meaningfully in nuclear research; this, in turn, creates a strong sense of ownership of processes as well as outcomes. Pham Ngoc Son and Bach Nhu Nguyen (Chapter 4) discuss the results obtained by the MNCP code to evaluate the correction factors for neutron activation foils.

The second section of the book is dedicated to neutron activation systems and technologies. Chapter 5 by Vitaly Krasilnikov, Mun Seong Cheon, and Luciano Bertalot provides a fascinating presentation of the neutron activation system (NAS), which was designed for ITER Tokamak. The main goal of the ITER NAS was to evaluate the total plasma neutron production rate. This system is under development by the Korean Domestic Agency of ITER.

Chapter 6 by Maria Angela de B. C. Menezes and Radojko Jaćimović provides an overview of the establishment of the methodology to analyze up to 5 g samples by k_0-instrumental NAA at CDTN, Brazil.

The editor wishes to thank all the authors for their high-level and valuable chapters. I would like to pay tribute to prominent scientists Profs. Nicolas Spirou and Andras Simonits, widely known for their research in the field of nuclear physics, who passed away last year. I would also like to thank Prof. G. Kennedy for his help concerning the drafting of several parts in the first chapter.

Thanks are due to Mr. Merzak Remki, manager at the Algerian Atomic Energy Commission COMENA. I am pleased to highlight his support promoting research and development.

I am also grateful to IntechOpen book manager Ms. Iva Lipović for her patience and assistance during the different edition phases of this book.

Finally, I would like to thank my husband Alghem Abdeldjalil for his help and encouragement during all my professional career and the preparation and edition of this book.

AQ: Please check if edits to the sentence beginning "Concentrations of trace levels..." retain the intended sense.

Dr. Lylia Alghem Hamidatou
Director of Nuclear Techniques and Applications Division
Nuclear Research Centre of Birine
Djelfa, Algeria

Section 1

Recent Applications of Neutron Activation Analysis

Chapter 1

Overview of Neutron Activation Analysis

Lylia Alghem Hamidatou

Additional information is available at the end of the chapter

http://dx.doi.org/10.5772/intechopen.85461

Abstract

This chapter provides a comprehensive overview of physical principles, procedures, proprieties, and some scientific achievements of neutron activation analysis. The most scientific events organized by the International Scientific Committees ICAA and k_0-ISC are also reported.

Keywords: neutron activation analysis, research reactor, gamma spectrometry systems, medicine and geology applications, advanced technologies, innovation

1. Principles

Neutron activation analysis is a physical technique that is based on nuclear reactions. The sample becomes radioactive when neutrons react with the nuclei of the elements' atoms. Radionuclides are formed and subsequently decay by emitting gamma rays that are unique in half-life and energy. Gamma-ray intensity is proportional to the element content in the sample. Instrumental neutron activation analysis (INAA) and k_0-INAA are the most sensitive analytical techniques used for the quantitative multielement analysis of major, minor, and trace elements in samples from almost every conceivable field of scientific or technical interest [1–9].

2. Procedures

The elements in a sample to be analyzed are made radioactive by irradiation with neutrons, and the induced radioactive samples can be identified and measured. The amount of a given neutron activation product that formed during neutron irradiation is proportional to the amount of its parent isotope and thus the concentration of the corresponding element [2].

3. Properties

The analytical techniques based on neutron activation are classified according to energy of incoming neutrons, elastic and inelastic neutron scattering, and neutron capture reactions as well as its combination such as thermal NAA (TNAA), epithermal NAA (ENAA), prompt gamma NAA (PGNAA), fast NAA (FNAA), and 14-MeV INAA a form of FNAA based on reactions with 14-MeV neutrons that are produced by neutron generators. These techniques have big application potential since they could provide data about large number of elements simultaneously. **Figure 1** presents more than 65 elements that may be analyzed using neutron activation analysis method. For most elements, NAA is an extremely sensitive method for analysis. The potential benefits of NAA include the multielement capability and the fact that non-destructive analyses can be performed.

4. Organization and scientific achievements

4.1. ICAA, k_0-ISC, and scientific events

The objectives of the International Committee on Activation analysis of ICAA are to improve and promote the application of nuclear methods of chemical analysis. The International Committee on Activation Analysis (ICAA) is organized exclusively for scientific and educational purposes, and specifically to coordinate the series of international conferences entitled Modern Trends in Activation Analysis (MTAA). The official ICAA website (http://www.icaa-mtaa.org) provides information on their activities, membership, etc. [10].

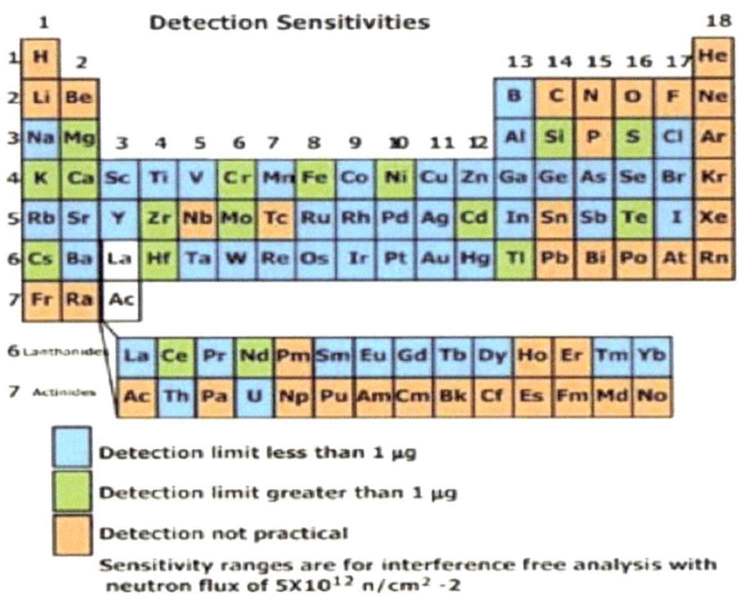

Figure 1. Detection sensitivities.

The k_0-International Scientific Committee was originally mainly involved in the organization of the k_0-Workshops and the accepting and reviewing of the papers presented there. The mandate of the k_0-ISC is to promote the development and application of the k_0-method of neutron activation analysis, k_0-NAA. Examples of the k_0-ISC's promotion of the development and application of the k_0 method include the coordination of the k_0-Workshops, maintenance and improvement of the recommended k_0-Nuclear Database, through the k_0-Nuclear Data Subcommittee, and the maintenance of the k_0 website to give information and to allow k_0-users to contact members of the k_0-ISC.

George de Hevesy (1885–1966) received the Nobel Prize for Chemistry in 1943 for his work on the use of isotopes as tracers in the study of chemical processes. The Hevesy Medal as illustrated in **Figure 2** is the premier international award of excellence honoring outstanding achievements in radioanalytical and nuclear chemistry as illustrated. **Table 1** presents the list of the laureates of George de Hevesy award during the period 1968–2018 [10].

4.2. International k_0-Users Workshops

The development of k0-NAA method is one of the most remarkable advances in the history of neutron activation analysis (**Table 2**). The k_0-NAA method includes comprehensive and accurate models of the neutron activation, radionuclide decay, and gamma-ray detection processes. During the period 1970–1980, Prof F. De Corte (University of Gent, Belgium) and Dr. A. Simonits (KFKI-AEKI, Budapest, Hungary) and many co-workers developed the concepts and procedures of the k_0-NAA method [4, 11–18]. Currently, this method is an inactive use in numerous laboratories all over the world [19–29]. The official k_0-ISC website (http://www.kayzero.com/k0naa/k0naaorg/k0-ISC.html) contains relevant information on the k_0-method and all associated events and database [24].

Figure 2. *George de Hevesy* (1885–1966) who received the Nobel Prize for Chemistry in 1943.

Year	Laureates	Year	Laureates
1968	W. Weine Meinke	2003	Jeroen J.M. De Goeij
1969	Albert A. Smales	2004	Attila Vértes
1970	Ivan Pavlovich Alimarin	2005	Zhifang Chai, Gregory Choppin, and Nicholas M. Spyrou
1972	Philippe Albert and Julian Hoste	2006	Jan Kucera
1975	Tibor Braun and Juraj Tölgyessy	2007	Robert R. Greenberg
1976	Francesco Girardi	2008	Syed M. Qaim
1977	Saadia Amiel Richard E. Wainerdi	2009	Richard M. Lindstrom
1978	Robert E. Jervis	2010	Darleane C. Hoffman
1979	Vincent P. Guinn	2011	Peter Bode (TU Delft page)
1981	William S. Lyon Max Peisach	2012	Boris F. Myasoedov
1983	Edward V. Sayre and Garman Harbottle	2013	Rajmund S. Dybczyński
1984	Georges Amsel	2014	Heino Nitsche
1985	Nobuo Suzuki	2015	Kattesh V. Katti and Susanta Lahiri
1986	Emile A. Schweikert	2016	Tomoko M. Nakanishi
2000	Frans De Corte	2017	Pavel P. Povinec
2001	Amares Chatt and Eiliv Steinnes	2018	Rolf Zeisler
2002	Enrico Sabbioni		

Table 1. List of the laureates of George de Hevesy award 1968–2018 [10].

4.3. Nuclear data

The authors F. De Corte and A. Simonits published in 2003, the recommended nuclear data for use in the k_0-standardization of neutron activation analysis [3]. In 2010, De Corte described in his paper "Towards an international authoritative system for coordination and management of a unique recommended k_0-NAA database" the constitution of the nuclear data library [25]. As indicated in the paper, "the 2012 recommended k_0 database", R. Javimovic et al. established the k_0 database in the form of an accessible excel file freely downloadable at http://www.kayzero.com/k0naa/News/News.html [26].

4.4. Neutron activation systems

Database from the IAEA's Research Reactor Database (RRDB) provides information with respect to the status of the world's research reactors. More than half of the 241 operational RRs worldwide performing neutron activation analysis distributed over 59 member states [31]. The highest roles of NAA have been identified as the most suitable opportunity for research, education and training, and commercialization of RR services. For that, NAA groups focused their efforts on the development and modernization of neutron activation analysis process including irradiation devices, gamma-ray spectrometers, and data analyzing instruments [31–35].

Title	Date	Organizer	References
MTAA-01	14–15 December 1961	Texas A&M College	
MTAA-02	18–21 April 1965	Texas A&M College	
MTAA-03	06–10 June 1968	National Bureau of Standards	NBS Spec Pub 312, June 1969, 2 vols
MTAA-04	01–05 October 1972	CNRS, Saclay, France	JRC v15–16, 1973 (also 17–19)
MTAA-05	13–17 September 1976	Institute of Radiochemistry, Technical University Munich, Germany	J. Radioanal. Chem., v. 37–39
MTAA-06	12 December 1981	University of Toronto, Canada	J. Radioanal. Chem., v. 69–72
MTAA-07	23–27 June 1986	Risø National Laboratory, Copenhagen, Denmark	J. Radioanal. Chem., v. 112–114
MTAA-08	30 December 1991	Technical University, Vienna, Austria	J. Radioanal. Chem., v. 167–169
MTAA-09	24–28 September 1995	Seoul, Korea	J. Radioanal. Nucl. Chem., v. 215–217 (1997)
MTAA-10	18–22 April 1999	National Institute of Standards and Technology (NIST)	J. Radioanal. Nucl. Chem., v. 244–245 (2000)
MTAA-11	20–24 June 2004	University of Surrey, Guildford, UK	J. Radioanal. Nucl. Chem., v. 271, No. 1–4 (2007)
MTAA-12	15–20 September 2007	Tokyo Metropolitan University, Hachioji, Tokyo, Japan	J. Radioanal. Nucl. Chem., v. 278, Issue 3, December 2008
MTAA-13	13–18 March 2011	Center for Chemical Characterization and Analysis Texas ASM University, College Station, TX, USA	J. Radioanal. Nucl. Chem., v. 291, Issue 2, February 2012
MTAA-14 and NAMLS-11	23–28 August 2015	Delft University of Technology, Delft, The Netherlands	J. Radioanal. Nucl. Chem., v. 309, Issue 1, July 2016

Table 2. List of the past international conferences on modern trends in activation analysis, MTAA (1961–2015) [10].

5. Innovation in NAA applications

Interesting paper published by P. Bode, the opportunities for innovation in NAA gives an overview by focused position-sensitive detection of elements in large samples, Monte-Carlo calculations replacing the use of standards, use of scintillator detectors, and new deconvolution techniques for increasing the sensitivity are examples of challenging new roads in NAA [30].

Menezes et al. reported in the paper "Advances in neutron activation analysis of large objects LSNAA with emphasis on archaeological examples", the recent developments and perspectives about the implementation of LSNAA and analyzing several kinds of matrices such as archeology, geology, and art objects [31]. During 2010–2015 period, the proficiency Tests and Inter-laboratory Comparisons have been carried out at the international level, by many NAA laboratories, under the framework of IAEA projects using IPE (International Plant-analytical Exchange) and ISE (International Soil-analytical Exchange) samples provided by the accredited organism such WEPAL [32].

Neutron activation analysis has traditionally been used mainly for the characterization of geological, environmental, and biological materials. However, other analysis techniques have emerged to replace NAA for many of those applications, and NAA now thrives mainly because of its unique advantage, the high penetrating power of neutrons and gamma rays, leading to ease of use in many instances where no sample preparation is required. In addition, NAA practitioners have innovated to provide fast, accurate, and reliable analyses of various matrices posing great difficulties for other techniques. Some of the applications made possible by these innovations are listed here.

In the medical field, it has often been hypothesized that a lack of selenium in the body may lead to increased risk for certain cancers. To measure the body's selenium status, toenails have been found to be a good indicator of the selenium status over a period of several months prior to the taking of the sample. Several NAA laboratories [37–39] have measured Se in toenails using the very short-lived Se-77m, half-life 17.5 s. This required the development of fast pneumatic sample transfer systems with accurate control of irradiation and decay times as well as accurate correction of counting losses when the count-rate is changing during the counting period. The results of studies [37–39] using thousands of samples sometimes revealed an association between lower Se levels and increased cancer risk, while in others, no significant difference in Se level was observed between the controls and subjects who developed certain forms of cancer.

A knowledge of the levels of trace elements in plastics may be useful from several points of view, as it may reveal the presence of toxic elements such as Cd [36], or information on the production process through the presence of catalyst residues, it may permit differentiating new plastics from those containing recycled material with flame retardants Br and Sb, and it may help decide the suitability of plastics for insulation of high-voltage electrical cables [37] or cables used in the nuclear industry [38]. Plastic samples of 100 mg mass are appropriate for some applications but several NAA laboratories have innovated to be able to analyze routinely and quickly samples up to 4 g mass which, for certain trace elements, are more representative of the original material, and in special cases, even larger samples may be analyzed using large-sample NAA.

The problem of the growth of fungus or mold on wood or paper products used in humid environments, wood for windows, cardboard on wallboard used in basements, and wrappers for bars of soap stored in bathrooms, has been remediated by the application of fungicides containing heavy elements like copper and iodine. Regular quality control measurements are needed to ensure that the right amount of fungicide has been applied. NAA is ideally suited for this as there is no sample preparation and methods have been developed [39] for large representative samples, at least 6 cm^2 for wood samples cut from the surface of the board and 60 cm^2 for paper, and the NAA results are independent of penetration depth. To successfully provide a fast and reliable service for industry, the reactor and staff must be available when needed; the service must be given high priority.

Pure selenium used for photographic film must be very low in chlorine; an upper limit of 2 μg/g is tolerable. Achieving this sensitivity by INAA is difficult because Cl is detected by 37 min half-life Cl-38 with gamma rays at 1642 and 2167 keV which suffer interference from the much more intense low-energy gamma rays emitted by several short-lived Se isotopes produced by neutron irradiation of Se. In order to reduce the detection efficiency for the interfering low-energy gamma rays relative to the high-energy Cl-38 gamma rays, discriminating gamma-ray spectrometry [40] has been used. A 10 mm thick lead plate was placed between sample and detector. To reduce the uncertainty of the calculated area of the barely visible 2167 keV peak, samples were counted for 30 m at the maximum tolerable count rate, which gave between 10 and 30% dead-time, and the peak fit was carried out using maximum predetermined information: the peak position and width were fixed relative to nearby strong Se peaks. A similar discriminating gamma-ray spectrometry technique was used [41] for the determination of vanadium in materials with high content of titanium, barium, and strontium.

To satisfy the increasing demand for rare-earth elements used in the electronics industry, new rare-earth mines are opening along with their associated refineries and the development of refining techniques. This development requires the measurement of rare-earth concentrations in the ore and products at all stages of refinement. The great sensitivity of NAA for most rare-earth elements has made it an excellent method for measuring rare-earths at low concentrations in rocks, sediments, soils, and plants. However, at the high concentrations found in ores and refinery products, the high neutron absorption cross-sections cause a severe neutron self-shielding problem even with small samples. This problem has recently been solved [42] with the development of an accurate neutron self-shielding model coupled with accurate gamma attenuation calculations for the low energy gamma rays of Ce, Nd, Sm, Dy, Ho, and Tm. The correction calculation is iterative, using the raw NAA-measured rare-earth concentrations in each sample as input and then simultaneously adjusting the neutron self-shielding factors and the gamma attenuation coefficients according to the models. For 100 mg samples with very high rare-earth concentrations, NAA results accurate to better than 5% were achieved for 13 rare-earth elements [43].

The use of nuclear analytical techniques is most practical to analyze a variety of samples in different fields related to life sciences in particular food product for humans. The objective of this research is to point out the research results of content and nutritional importance of individual essential elements that are present in various milks and dairy products. Recently, an extensively studied on nutrients and heavy metals concentration in food overall the world. In the last years, several brands of milk powder for categories child and adult have been studied by using the k_0-INAA technique [44]. As complementary work for dairy products, it is very important to evaluate the consumption of the commercial baby formula milk for the first and the second age in Algeria. The chemical element Ca, K, Na, Br, Rb, and Zn have been determined in six kinds of baby formula milk using k0-NAA technique. Results obtained in this work show a good agreement with concentration values given by producers [9].

Relevance of activation analysis is driven by its stakeholders. As an example of the existing stakeholder's needs are mainly to evaluate the whole blood and hair correlation factor of Na and K, Fe, and Zn of patients suffering from Alzheimer and psoriasis pathologies and normal controls using INAA and k_0-NAA techniques. According to the gender and age consideration, the results obtained in these studies have been discussed [45, 46]. For quality control and quality assurance, the biological matrix NIST1566b Ouster Tissue), IAEA A13 (animal blood) were analyzed simultaneously with the samples. Three statistical parameters Z-score, U-score, and bias were determined and discussed.

For more details about the basic theory of NAA, recent developments and technologies and potential applications of neutron activation analysis performed at the author's laboratory and nuclear facilities, the interested reader should refer to the works discussed in Chapters 2–6.

Chapter 2: Neutron Activation Analysis: Application in Geology Application in Geology and Medicine (*Maitreyee Nandy*).

Chapter 3: Colombian Neutron Activation Analysis Laboratory – CNAAL. Applications and Development using the Nuclear Research Reactor IAN-R1 (*Guillermo Parrado et al.*).

Chapter 4: Monte Carlo simulation of Correction Factors for Neutron Activation Foils (*Pham Ngoc Son*).

Chapter 5: Neutron Activation System for ITER Tokamak (*Vitaly Krasilnikov et al.*).

Chapter 6: An Overview of the Establishment of Methodology to Analyze up to 5 g-Sample by k_0-Instrumental Neutron Activation Analysis, at CDTN, Brazil (*Maria Menezes et al.*).

Acknowledgements

The author wants to express her deep conviction and my gratitude with regard to the role promoting nuclear research and development, one of such importance, which Mr M. Remki, the Top Manager of the Algerian Atomic Energy Commission is playing. The author would like to thank Prof. G. Kennedy retired from the Engineering Physics Department, Polytechnique, Montreal, Canada for his comments and suggestions. I am very grateful to my colleagues at the Nuclear Techniques and Applications Division, CRNB, COMENA.

Author details

Lylia Alghem Hamidatou

Address all correspondence to: lylia.h@hotmail.fr

Nuclear Research Centre of Birine, Algerian Atomic Energy Commission, CRNB/COMENA, Algeria

References

[1] Glascock MD. Neutron activation. In: Alfassi ZB, editor. Instrumental Multi-Element Chemical Analysis. Dordrecht, the Netherlands: Kluwer Academic Publishers; 1998. pp. 93-150

[2] Hamidatou L, Slamene H, Akhal T, Zouranen B. Concepts, instrumentation and techniques of neutron activation analysis. In: Kharfi F, editor. Imaging and Radioanalytical Techniques in Interdisciplinary Research. IntechOpen. Available from: https://www.

intechopen.com/books/imaging-and-radioanalytical-techniques-in-interdisciplinary-research-fundamentals-and-cutting-edge-applications/concepts-instrumentation-and-techniques-of-neutron-activation-analysis

[3] Alghem L, Ramdhane M, Khaled S, Akhal T. The development and application of k_0-standardization method of neutron activation analysis at Es-Salam research reactor. Nuclear Instruments and Methods. 2006;**556**:386-390

[4] Hamidatou LA, Ramdhane M. Characterization of neutron spectrum at Es-Salam research reactor using Høgdahl convention and Westcott formalism for the k_0-based neutron activation analysis. Journal of Radioanalytical and Nuclear Chemistry. 2008;**278**(3):627-630

[5] Hamidatou L, Benkharfia H. Experimental and MCNP calculations of neutron flux parameters in irradiation channel at Es-Salam reactor. Journal of Radioanalytical and Nuclear Chemistry. 2011;**287**(3):971-975

[6] Hamidatou LA, Dakar S, Boukari S. k_0-NAA quality assessment in an Algerian laboratory by analysis of SMELS and four IAEA reference materials using Es-Salam research reactor. Nuclear Instruments and Methods. 2012;**682**:75-78

[7] Khaled S, Belaid S, Mouzai M, Alghem L, Ararem A, Zouranen B. Development and application of the cyclic delayed neutron counting technique at Es-Salam research reactor. Physical and Chemical News. 2009;**45**:39-43

[8] Hamidatou L, Slamene H, Akhal T, Begaa S, Messaoudi M, Boussaad Z. NAA Algerian laboratory evaluation processed by WEPAL and IAEA during 2011-2012. Journal of Analytical Sciences, Methods and Instrumentation. 2013;**3**:184-192

[9] Hamidatou L, Slamene H, Akhal T, Boulegane A. Trace and essential elements determination in baby formulas milk by INAA and k_0-INAA techniques. Journal of Radioanalytical and Nuclear Chemistry. 2014;**301**(3):659-666

[10] International Committee on Activation Analysis. Available from: http://www.icaa-mtaa.org

[11] De Corte F. The k Standardization Method—A Move to the Optimization of NAA. University of Gent; 1987

[12] Dung HM, Hien PD. The application and development of k_0-standardization method of neutron activation analysis at Dalat research reactor. Journal of Radioanalytical Chemistry. 2003;**257**:643-647

[13] Jacimovic R. Aggregate [thesis]. University of Ljubljana; 2003

[14] Dung HM, Sasajima F. Determination of alpha and f for k_0-NAA in irradiation sites with high thermalized neutrons. Journal of Radioanalytical Chemistry. 2003;**257**:509-512

[15] Dung HM, Cho SY. Journal of Radioanalytical Chemistry. 2003;**257**:573

[16] Lin X, Baumgartner F, Li X. Journal of Radioanalytical Chemistry. 1997;**215**:179

[17] Rizo OD, Peraza EH, Reyes MCL, Pellon IA, Guevara MVM, Cabrera MI. Journal of Radioanalytical Chemistry. 1997;**220**:95

[18] Acharya R, Chat A. Characterization of the Dalhousie University SLOWPOKE-2 reactor for k_0-NAA and application to medium-lived nuclides. Journal of Radioanalytical and Nuclear Chemistry. 2003;**257**:525-529

[19] Chung YS, Dung HM, Moon JH, Park KW, Kim HR. Nuclear Instruments and Methods A. 2006;**564**:702

[20] Soing WB, Dung HM, Wood AK, A-Salim NA, Elias Md S. Nuclear Instruments and Methods A. 2006;**556**:716

[21] Khoo KS, Sarmani SB, Abugassa IO. Determination of thermal to epithermal neutron flux ratio (f), epithermal neutron flux shape factor and comparator factor (Fc) in the Triga Mark II reactor, Malaysia. Journal of Radioanalytical Chemistry. 2007;**271**:419-424

[22] Jacimovic R, Smodis B, Bucar T, Stegnar P. Journal of Radioanalytical and Nuclear Chemistry. 2003;**257**:659

[23] De Corte F, Simonits A. Recommended nuclear data for use in the k_0-standardization of neutron activation analysis. Atomic Data and Nuclear Data Tables. 2003;**85**:47-67

[24] Available from: http://www.kayzero.com/k0naa/k0naaorg/k0-ISC.html

[25] De Corte F. Towards an international authoritative system for coordination and management of unique recommended k_0-NAA database. Nuclear Instruments and Methods A. 2010;**622**:373-376

[26] Jacimovic R, De Corte F, Kennedy G, Vermearcke P, Revay Z. Journal of Radioanalytical and Nuclear Chemistry. 2014;**300**:589-592

[27] Training Workshop on the IAEA Neutron Activation Analysis E-learning Course IAEA Headquarters; Austria 3-7 September 2018; Vienna

[28] The regional training course on quality assurance and quality control in analytical and radio-analytical laboratories as the RAF7015 project. In: Strengthening Regional Capacities for Marine Risk Assessment Using Nuclear and Related Techniques; 23-27 October 2016; Algiers, Algeria

[29] IAEA-TECDOC-1839: 978-92-0-103818-0. Development of an Integrated Approach to Routine Automation of Neutron Activation Analysis. 2018

[30] Bode P. Opportunities for innovation in neutron activation analysis. Journal of Radioanalytical and Nuclear Chemistry. 2012;**291**(2):275-280

[31] Menezes MABC, Jaćimović R, Ribeiro L. In: International Atomic Energy Agency, editor. Contribution of analytical nuclear techniques in the reconstruction of the Brazilian prehistory analysing archaeological ceramics of Tupiguarani tradition, Advances in Neutron Activation Analysis Of Large Objects with Emphasis on Archaeological Examples. Research Project. IAEA-TECDOC-1838. 2018. pp. 4-21

[32] IAEA-TECDOC-1831: 978-92-0-108617-4. Proficiency Testing by Inter-laboratory Comparison Performed in 2010-2015 for Neutron Activation Analysis and Other Analytical Techniques. 2017

[33] Van den Brandt PA, Goldbohm RA, van't Veer P, Bode P, Dorant E. A prospective cohort study on selenium status and the risk of lung cancer. Cancer Research. 1993;**53**:4860-4865

[34] Garland M, Morris JS, Stampfer MJ, Colditz GA, Spate VL, Baskett CK, et al. Prospective study of toenail selenium levels and cancer among women. Journal of the National Cancer Institute. 1995;**87**:497-505

[35] Ghadirian P, Maisonneuve P, Perret C, Kennedy G, Boyle P, Krewski D, et al. A case-control study of toenail selenium and cancer of the breast, colon, and prostate. Cancer Detection and Prevention. 2000;**24**:305-313

[36] Bode P. The use of INAA for the determination of trace elements, in particular cadmium, in plastics in relation to the enforcement of pollution standards. Journal of Radioanalytical and Nuclear Chemistry. 1993;**167**:361-367

[37] Crine JP, Pelissou S, St-Onge H, St-Pierre J, Kennedy G, Houdayer A, Hinrichsen P. Elemental and ionic impurities in cable insulation and shields. In: Proc. Jicabel 87 Conf.; 1987. pp. 206-213

[38] Kucera J, Cabalka M, Ferencei J, Kubešová M, Strunga V. Determination of elemental impurities in polymer materials of electrical cables for use in safety systems of nuclear power plants and for data transfer in the Large Hadron Collider by instrumental neutron activation analysis. Journal of Radioanalytical and Nuclear Chemistry. 2016;**309**:1341-1348

[39] Kennedy G. Treating wood with chemical preservatives. In: Analytical Applications of Nuclear Techniques. Vienna: IAEA; 2004. pp. 123-125. Available from: http://www-pub.iaea.org/MTCD/publications/PDF/Pub1181_web.pdf

[40] Kennedy G, Zikovsky L. Improvement of sensitivity in neutron activation analysis by selective absorption of high-intensity low-energy gamma-rays. Journal of Radioanalytical Chemistry. 1982;**72**:295-304

[41] Kamenik J, Dragounova K, Kucera J, Bryknar Z, Trepakov VA, Strunga V. Determination of vanadium in titanate-based ferroelectrics by INAA with discriminating gamma-ray spectrometry. Journal of Radioanalytical and Nuclear Chemistry. 2017;**311**:1333-1338

[42] Chilian C, Lacroix C. Towards routine NAA of materials rich in heavy elements with iterative gamma-ray attenuation and neutron self-shielding calculations. Journal of Radioanalytical and Nuclear Chemistry. 2014;**300**:547-552

[43] Abdollahi Neisiani M, Latifia M, Chaouki J, Chilian C. Novel approach in k_0-NAA for highly concentrated REE Samples. Talanta. 2018;**180**:403-409

[44] Hamidatou LA, Khaled S, Mouzai M, Zouranen B, Ararem A, Alghem A, et al. Instrumental neutron activation analysis of milk samples using the k_0-standardization method at Es-Salam research reactor. Physical and Chemical News. 2009;**45**:44-47

[45] Mansouri A, Hamidatou Alghem L, Beladel B, Mokhtari OEK, Bendaas A, Benamar MEA. Hair-zinc levels determination in Algerian psoriatics using Instrumental Neutron Activation Analysis (INAA). The International Journal of Applied Radiation and Isotopes. 2013;**72**:177-181

[46] Doumaz T, Hamidatou L, Beladel B, Slamene H, Begaa S, Messaoudi M, et al. Determination of Na, K, Fe, and Zn whole blood elements concentration of Algerian Alzheimer patients by k_0-NAA method. Journal of Radioanalytical and Nuclear Chemistry. 2016;**309**:229-234. DOI: 10.1007/s10967-016-4760-2

Chapter 2

Neutron Activation Analysis: Application in Geology and Medicine

Maitreyee Nandy

Additional information is available at the end of the chapter

http://dx.doi.org/10.5772/intechopen.76726

Abstract

Varied forms of neutron activation analysis (NAA), due to their high accuracy and reproducibility, are being used in geological studies and in medical application for the determination of concentration of elements down to the trace and ultra-trace level. Concentration of Cs, Sc, Fe, Ta, Co and Eu which may give rise to long-lived activity on neutron irradiation has been determined down to 0.1 ppm in rock samples from 11 geological formation in Karnataka, India, using NAA. NAA has been used by several authors to determine elemental concentration in biological shields, different geological formation around the world, thermal springs, archaeological objects and precious stones. NAA has been successfully employed by different groups to determine the concentration of Al, K, Na, Cl, Rb, Ca, Cu, Co, I, Mg, Se, Fe, Zn, Hg, Ba, Cr, etc. and their relative variation in breast cancer, skin cancer, colorectal cancer, dysfunction and malignancy of thyroid gland.

Keywords: neutron activation analysis (NAA), elemental concentration, geology, medicine, radioactivity

1. Introduction

Neutron activation analysis (NAA) is a state-of-the-art analytical technique used for identification of elements and determination of the elemental concentration in different applications. With the development of accurate and fast gamma spectrometric techniques, advanced electronics and automation in data analysis, the method has gained active application in diverse field of basic research. Non-destructive or instrumental NAA has also become popular for varied industrial application due to its ease in handling. With the development of high-power neutron source, precise and fast neutron detectors and sophisticated electronics, NAA has achieved ultralow minimum detection limit (MDL). As a result the method has turned out to be one of the best choices for quantification of trace elements in a wide variety of samples—geological, archaeological, biological and environmental samples.

Varied analysis devices for neutron activation studies have made it a preferred choice for the analysis of construction material, coal, environmental, geological and archaeological samples and biological material, security monitoring and academic studies. Since its inception around the middle of last century, different variants of NAA such as prompt gamma neutron activation analysis (PGNAA), pulsed fast neutron analysis (PFNA), pulsed fast/thermal neutron analysis (PFTNA) and associated particle imaging (API) have been developed. Non-destructive NAA is one of the most preferred methods of sample analysis as it requires minimal processing of the activated sample. In the semiconductor industry, NAA is used to determine ultra-trace concentration of impurities, or dopant as impurity concentration even at 1 ppb can affect the performance of the semiconductor. NAA-induced associated particle imaging (API) efficiently used for security monitoring has several advantages over other conventional scanning methods.

Different forms of NAA offer good choice over other techniques for application in geology and medicine. NAA is used for quantitative analysis of trace elements in rock samples which serves as an important tool for modelling geochemical processes and also helps in sample selection for other applications. Pai et al. [1] have used instrumental NAA (INAA) for trace element analysis in different rock samples to determine its suitability for use in the coarse aggregate for concrete in shielding material. Over the last several years, NAA has made important inroads in the field of medical science. The method is used in production of radiotracers which are used in situ for evaluation of new pharmaceuticals for their distribution, time release, clearance, etc. Neutron-stimulated emission computed tomography (NSECT) is a newly developed imaging method that non-invasively maps the concentration of an isotope in the body [2]. Multiple pencil beams of fast neutrons are developed for early diagnosis of breast cancer through NAA using differential femto-oximetry (DFO) [3]. This chapter will discuss application of different forms of NAA in geology and medicine and related recent advancements.

2. Method

Neutron activation analysis (NAA) is a procedure employed for analysing the elemental composition of a material and for determining the concentrations of elements at the trace and ultra-trace level in a vast majority of samples. In this method the sample is irradiated with neutron from the reactor when nuclear reaction is induced in the sample. The gamma rays emitted from the isotope as a result of this reaction is detected and counted and the reaction analysed. The activity of the ith isotope induced in the sample is given by

$$A_i = N\left[\int \sigma_i(E)\varphi(E)dE\right]\left(1 - e^{\lambda_i t_I}\right) \tag{1}$$

where N is the number of target atoms available for interaction, σ_i is the cross section for production of the ith isotope, $\varphi(E)$ is the neutron fluence rate at energy E, λ_i is the decay constant for the ith isotope, t_I is the time of irradiation.

Gamma spectrometric analysis of the irradiated sample helps one to determine the elemental composition and impurity of the sample. There are different types of NAA that can be used for the analysis of a sample.

In many applications the sample under study is irradiated in a reactor or by a neutron source, and the exposed sample is analysed for gamma emission. No chemical treatment of the sample is carried out between irradiation and counting. This type of NAA is called instrumental neutron activation analysis or INAA. This is also known as non-destructive neutron activation analysis. In the second method, the sample after irradiation with neutrons is treated chemically in order to separate different constituents depending on their solubility or other physical properties. The separated constituents are then counted separately for their characteristic gamma emission. This method is called radiochemical neutron activation analysis (RNAA). INAA can again be carried out in more than one way—prompt gamma neutron activation analysis (PGNAA) and delayed gamma neutron activation analysis (DGNAA). In the first method, gamma counting is done online when the sample is irradiated. The neutron interacts with the target nucleus, and the latter goes to an excited state which thereby decays to the lower energy states with the emission of gamma radiation. The reaction time is of the order of 10^{-15} s. Analysis of these gamma rays provides information on the incident neutron flux as well as the composition of the sample. In DGNAA gamma spectrometry of the irradiated sample is carried out after the irradiation is over. In this method the decay gammas from the product radioisotopes are counted. The k0-NAA is one of the most accepted standardisation methods for quantitative multielemental analysis of the sample through gamma spectrometry. In this method one standard or reference material with known amount of an isotope and a single comparator are irradiated and counted along with the sample under the identical conditions. The quantity of the different elements present in the sample is estimated using the k-factors determined from the analysis of the standard and the comparator. The varied forms of NAA are applied for quantitative analysis of different types of samples.

3. Neutron activation in geology

Different variant of NAA is employed for detection and quantification of the trace and ultra-trace level of elemental concentration in different geological formation, uranium ore, etc.

NDNAA and Compton-suppressed NAA are efficiently used for analysis of geological samples (i) for uranium content measurement in phosphate rocks and surrounding geology and (ii) in trace elemental analysis of rock samples. Trace elemental analysis of rock samples is particularly important as these rocks can find a novel use as the coarse aggregate in varieties of concrete that may be developed for shielding of accelerator and reactor facilities [1].

Analysis of rare earth and other trace elements in the rock samples helps geoscientists to understand the chemistry of rock formation [4]. NAA studies also suggest that extinction of the dinosaurs occurred soon after the impact of a large meteorite with the earth.

Neutron reaction cross section strongly depends on the energy and sharply increases as energy decreases below 0.5 eV (**Table 1**).

Phosphate rocks contain other elements adjacent in the periodic table. Of these are Na, Mn and Cl which may interfere with the detection of uranium. In the following portion, we shall have a look on the gamma spectrometric properties of the stable isotopes of these elements (**Table 2**).

Precise determination of the elemental content of the phosphate rocks is crucial in the understanding of geochemistry of the formation. INAA and delayed neutron activation analysis (DNAA) were employed to determine the concentration of uranium in some Egyptian environmental samples (Toshki soil, Aswan iron ore) and phosphate samples in the Red Sea coast area [4]. The study showed, in consonance with studies by other workers, that the phosphate rocks are rich in natural sources of uranium.

INAA coupled with gamma ray spectrometry was employed to determine the concentration of the three major elements U, Th and K collected from the lower Benue region of Nigeria [5]. The authors observed a higher abundance of Th than in the other regions. For some samples as the potassium contents were fairly normal, high values of the Th (ppm)/K ratios were obtained.

Shutdown of reactors, accelerators and other radiation facilities leads to generation of radioactive waste from shielding and other materials of the facility which got activated over the years

^{238}U	Thermal, 2.683 b; epithermal, ~1 b with sharp resonances
^{235}U	Thermal, 98.8 b; epithermal, ~10 b with sharp resonances

Table 1. Neutron capture cross sections for ^{238}U and ^{235}U.

Stable isotope	Cross section (n, γ) (b)	Half-life (radioisotope)	Gamma energy (MeV) (intensity)
	Thermal		
^{23}Na	0.4	14.96 h (^{24}Na)	1.37 (99.99)
			2.75 (99.93)
^{37}Cl	0.4	37.23 min (^{38}Cl)	1.64 (33.3)
			2.16 (44.4)
^{55}Mn	15	2.58 h (^{56}Mn)	0.847 (98.85)
			1.81 (26.9)
			2.11 (14.2)
^{81}Br	0.25	1.27 min (^{82}Br)	0.511 (21.0)
			0.776 (99.0)

Table 2. Gamma spectrometric properties of isotopes relevant for INAA of phosphate rock.

http://dx.doi.org/10.5772/intechopen.76726

of operation of the facility. Radioactive concrete is one of the major constituent of this waste. The concrete shield in reactors, low- and medium-energy accelerator facilities contain induced activity from neutron activation, while in high-energy accelerators, gamma activation also plays a significant part. Long-lived radioactivity in the concrete waste poses a serious problem for its management and disposal. This entails a detailed in-depth analysis of the elemental composition of the concrete to be used in the shielding of the accelerator facilities.

Coarse and fine aggregates are the primary components of concrete. In their work for development of self-compacting concrete (SCC), Pai et al. [1] had analysed rock samples from 11 different geological formation from the state of Karnataka, India. The different types of rocks analysed were gneiss, granite, trap, basalt, peninsular gneiss, dolomite rock, Deccan trap, sandstone, quartzite, limestone and laterite. Elemental composition of the samples was determined using non-destructive neutron activation followed by low-background gamma spectrometry. Concentration of the elements Cs, Sc, Fe, Ta, Co and Eu which may result in long-lived activity build-up was measured down to 0.1 ppm. A detailed analysis of the activated samples from all the 11 geological formation showed that quartzite has the lowest concentration of these elements and hence has the minimum activity build-up [1]. But it failed the fresh and hardened property test which determines the acceptance criteria for the SCC. Hence, dolomite rock which showed the second lowest activity build-up was inferred as the most suitable choice to be used as the coarse aggregate in developing the SCC. Some other types of rocks like Deccan trap and laterite showed a high concentration of neutron-induced activity and are not desirable during decommissioning or disposal. Hence, these are not recommended to be used as a constituent of the concrete shielding.

Medhat et al. [6] have assessed the contents of trace elements, in the biological shield of a decommissioned nuclear power reactor, which gives rise to significant amounts of radioactivity after long periods of operation. Several cement samples were studied through fast neutron irradiation followed by gamma spectrometry. The (n,γ) reaction was monitored to estimate the concentration of different elements like Ce, Co, Cs, Eu, Fe, Hf, Sb, Sc, Ta and Tb.

Nazarov et al. [7] studied the activity generation in the concrete of the shielding material of radiation facilities in order to understand the impact on their decommissioning procedure. Since the activity is produced mostly through (n,γ) reaction, analysis of the elemental composition of the raw materials plays a crucial role in the choice of the components of the cement and concrete. This elemental analysis was carried out by neutron activation and subsequent gamma spectrometric studies. The authors observed that the analysis of the elemental composition needs to be carried out at the design stage of the facility. This would give an idea of the radioactive waste generated at the decommissioning stage.

Alden et al. [8] has used INAA to determine the elemental composition of 157 archaeological samples and geological clay from Northern Chile. They observed that the major groups consisting of high chromium and low chromium were found in clays from within the region. The three minor groups containing low sodium comprised the vessels imported from northwest Argentina. The distribution pattern indicated that the constituent material differed for different groups of items. Compositional analyses also indicated that the Inka-style ceramics were produced in the region during the period of Inka domination.

Huckell et al. [9] has used INAA to determine the composition of microcrystalline sedimentary rock (chert) from the western part of North Dakota. The rocks came from a vast geological formation on the central and northern plains and were a part of the Eocene-age White River Group. They found that elemental composition of the rocks analysed was distinct and helped its separation from the other cherts.

Michelsen and Steinnes [10] had used thermal neutron activation to estimate the relative abundance of copper in some geological samples. The activated samples were analysed by gamma spectrometry using the annihilation photons from ^{64}Cu which is a positron emitter. The coincidence counting of annihilation photons largely reduces the interference from ^{24}Na. ^{24}Na can be formed in the sample through neutron activation of the stable isotope ^{23}Na which is often present in significant amount in geological materials. Copper concentration to the level of 100–1000 ppm could be successfully determined in this method.

Ravisankar et al. [11, 12] used instrumental neutron activation analysis to detect and determine the elemental composition of some beach rock samples in the South East Coast of Tamilnadu, India. Along with other constituents, concentrations of the rare earth elements (REE) were measured. The authors used single comparator method in this study. They determined the fraction of 19 elements in the 15 samples collected from the area. High-resolution gamma spectrometry was used for the purpose. Irradiated Standard Reference Material (SRM 1646a Estuarine sediment) was measured for calibration. The geochemical behaviour of REE in beach rock was studied.

Natural emeralds, associated rocks obtained from Rajasthan, India, were analysed for contents of different elements [13]. The content ratio for 21 elements was estimated by instrumental neutron activation analysis (k0 INAA method) and high-resolution gamma ray spectrometry. Elemental analysis of the samples carried out by the authors showed that some elements from the trapped and host rocks were separated and preferentially combines with the mineral beryl forming the gemstones. For quantitative analysis of the samples, the results were compared with those for the reference rock standard of the US Geological Survey (USGS BCR-1).

Several works using INAA were carried out by El-Taher A. and his group to characterise the elemental composition of different geological formation around the desert in Egypt and the Egyptian Red Sea coast. INAA of 25 elements, As, Ba, Ca, Ce, Co, Cr, Cs, Eu, Fe, Hf, K, La, Lu, Mg, Mn, Na, Nd, Rb, Sc, Sm, Th, U, Yb, Zn and Zr, was carried out in quartz collected from the region [14]. Activation of the samples was done in the TRIGA Mainz research reactor and the activated samples were analysed using standard gamma spectrometric method. As claimed by the authors, these results are the first data for elemental composition of Egyptian quartz.

Detailed quantitative analysis was carried out by the same group also for granite samples collected from different locations in South Egypt [15, 16]. Elemental content for a total of 28 elements comprising of several rare earth elements (REEs) was estimated. INAA as a sensitive non-destructive analytical tool is a preferred choice for quantitative determination of REEs because strong similarity in chemical behaviour of these elements renders the chemical methods inefficient for identification and estimation of the elements. The elements quantified were Na, Mg, K, Fe, Mn, Sc, Cr, Ti, Co, Zn, Ga, Rb, Zr, Nb, Sn, Ba, Cs, La, Ce, Nd, Sm, Eu, Yb,

Lu, Hf, Ta, Th and U. The results were compared with those obtained from X-ray fluorescence of the same samples. Next, in this series of work by El-Taher et al., concentration of chromium and 15 trace elements was measured in the chromite rock samples collected from the Eastern Desert in Egypt [17]. Short periods of irradiation were used to measure the short-lived isotopes of Mg, Ti and Mn, while longer irradiation periods were used for quantification of Na, Ga, As, La, Sc, Cr, Fe, Co, Zn, Zr, Ce, Ce, Yb, Lu, Hf and Ta.

El-Taher et al. have used INAA to estimate the concentration of gold and 31 other elements in the 10 samples collected from two Egyptian gold ores in the Eastern Desert region [18]. They determined 42.4% and 25.7% abundance of gold for the samples collected from the two ores.

Schwedt et al. employed neutron activation to analyse ceramic vessels found during excavations of ancient cemeteries in different parts of Boeotia, Greece [19]. The analyses resulted in a clear separation between the samples obtained from the shores of the ancient lake Copais and those from eastern parts. Analysis of the elemental compositions revealed that for some of the samples the results were similar to those obtained among Bronze Age samples from the same region. On the other hand, composition of the samples from the Theban tombs indicated import from regions as distant as Asia Minor.

Srivastava et al. [20] has quantified the concentrations of selenium, arsenic and 13 other elements in soil samples rich in selenium from the state of Punjab, India. The samples were irradiated in research reactor, and the activated samples were counted using high-resolution γ ray spectrometry. Quantification was done using the comparator method using soil samples collected from a non-seleniferous region.

Quantitative analysis of the composition of pumice obtained from archaeological excavations throws light on important parameters like their origin, age, transport route, etc. Steinhauser et al. have used neutron activation on pumice samples from the Mediterranean regions Milos, Santorini, Kos, Gyali and Nisyros (Greece), Lipari (Italy) and Cappadocia (Turkey) [21]. Through gamma spectrometric analysis of the activated samples, they have determined the concentrations of the elements As, Ba, Ce, Co, Cr, Cs, Eu, Fe, Hf, K, La, Lu, Na, Nd, Rb, Sb, Sc, Sm, Ta, Tb, Th, U, Yb, Zn and Zr and inferred about the origin and ages of the pumice samples.

Wasserman et al. [22] have studied the geochemistry of Sepetiba Bay in four sediment cores. INAA was used to determine the total concentration of the metals Ba, Co, Cr, Cs, Fe, Hf, Rb, Sc and Zn; the rare earth elements Ce, Eu, La, Lu, Sm and Yb; the actinides Th and U, along with As and Br; organic carbon; and total sulphur. They found that the top layers had strong zinc contamination (1000 μg/g or higher).

Instrumental neutron activation analysis combined with pattern recognition techniques was applied by Watterson et al. to understand the mineralisation process in granites and classification of diamonds and to identify the sedimentary units in Witwatersrand, South Africa [23]. Classification of coals from the Witbank Coalfield was also carried out. The method proved to be an efficient tool for identification of sedimentary unit mapping of geological formation.

Limestone samples from different areas in the Longmen Grottoes, Henan province, China, were analysed by Zhu et al. [24]. Concentrations of major elements, trace elements and REEs were determined. The authors identified three different groups of rocks—the samples from the northern part were made of dolomite rocks, and those from the middle and southern part are mainly made of limestone.

Contis [25] has measured the concentration of Se, V, As, Hg and Cd in the water from thermal springs using NAA. The study was carried out to understand the effect of thermal springs in the eastern Aegean Sea in Greece, Ikaria, on drinking water from sources near and far from the spring. The measured concentration of the elements did not show statistically significant variation from that due to natural background.

4. Neutron activation in medicine

NAA is used in different applications in medicine and has made inroads in the newer areas.

Trace elements have important bearing on the physiological and biochemical processes, and relative abundance and balance of different elements in trace quantities strongly influence the occurrence and advancement of many diseases. Estimation of trace element concentration in breast cancer, skin cancer, colorectal cancer, dysfunction and malignancy of thyroid gland has been done using neutron activation followed by high-resolution gamma spectrometric studies. NAA is one of the preferred choices for trace- and ultra-trace-level quantitative estimation as it is a highly accurate, precise and reproducible method even for measurements to the ppm and ppb level.

Concentration of Ca, Cu, Co, I, Mg, Se, Fe, Zn, Hg, Ba and Cr at the trace level in the malignant tissues of colorectal cancer was determined through NAA by H. Arriola et al. [26]. The study was carried out for patients from Mexican population. The results were compared with those obtained for normal tissues in the same population. It was observed that the amount of Co, Fe, I and Ba changes due to incidence of colorectal cancer.

NAA was used to estimate the concentration of Al, K, Na, CI, Rb, Co, Sc, Mn, Mg, Se, Zn, Cs, Fe and Cr in the patients of breast cancer in Sudanese population (Ammar Mubark Ebrahim A M, 2003) [27]. Though the number of patients studied was only 80, it was found that the amount of Al, Mg, Cr, Mn, Se and Zn is higher in the malignant tissues compared to the normal tissue. In this work the author has observed variation in the levels of K, Na, Fe, Co, Sc, Rb and Cs but to a much lesser extent.

Rees et al. [28] have used INAA to detect and measure the arsenic content in the toenail of the skin cancer patients. Tissues from both basal skin carcinoma (BCS) and squamous cell carcinoma (SCC) were examined from the patients of non-melanoma skin cancer.

Zaichick and Zaichick [29] have studied the changes in trace element concentration in the cancerous human prostate gland. They have used INAA to measure the concentration of 43 trace elements and compared the results for malignant, benign hypertrophic and normal prostate. Of the 43 elements measured, concentration of 33 elements in the malignant prostate is higher than that in the benign hypertrophic tissue. For the elements Co, Hg, Rb, Sc, Se and

Zn, a reverse trend was observed. When compared with the normal tissue, lower contents of Sc, Se, Zn, Rb and Cd were detected in the malignant prostate.

Zaichick et al. [30] have studied the role of trace elements in the induction and advancement of thyroid cancer. Contents of 11 elements, namely, Sc, Se, Zn, Co, Cr, Fe, Hg, I, Rb, Sb and Ag were determined in malignant and non-malignant thyroid nodules as well as in the non-affected paranodular thyroid tissue. Measured concentration of the elements mentioned was compared with the reference standard material H-4 of the International Atomic Energy Agency (IAEA). The results of this study showed that the level of Ag, Co, Hg, I and Rb is higher in the paranodular tissue. Selenium deficiency was also reported in this work for malignant thyroid. Zaichick and Zaichick [31] also studied the influence of different elements in the functional behaviour of thyroid and their dependence on sex and age.

Neutron activation analysis through monitoring of the delayed gamma radiation (DGNA) in combination with dual-energy X-ray absorption was used by Aloia et al. [32] to estimate the total body calcium. For the same population, the results obtained by the two methods vary more than 20%.

Gamma ray imaging plays a significant role in the pharmaceutical industry in development and progress of a drug delivery system. The standard form of radiolabelling of the drug molecules is done using some of the most commonly used medically important radioisotopes, like ^{99m}Tc or ^{111}In. But for complex drug molecules, radiolabel is produced through in situ neutron activation [33].

NSECT or neutron-stimulated emission computed tomography is one of the most advanced imaging techniques employed to study the isotope distribution in biological tissue [2, 34]. The method depends on irradiation of the sample by fast neutrons. The gamma rays emitted in the nuclear reaction induced in the isotopes in the tissue under study are monitored to construct tomographic images of each section of the sample. Though the instrumentation is expensive, high sensitivity of the technique has rendered it suitable to be used for cancer staging, detection of breast cancer.

Another important advancement in the realm of nuclear medicine is the early diagnosis of breast cancer with the help of the fast neutrons. Multiple pencil beams are developed to carry out NAA of the breast tissue. Since the oxygen content of the cancer tissue is different from that in the normal tissue, differential femto-oximetry is used in the diagnosis of the malignant tissue [3]. With the advancement in beam profile variation, the technique may be used for diagnosis of other types of cancer.

5. Conclusion

Neutron activation analysis is one of the preferred techniques for quantitative analysis of different types of samples and has thus found wide application. It is a highly accurate method and can reliably be applied for measurement of concentration of elements at the trace and ultra-trace level. We have discussed how NAA, particularly INAA, can be used for determination of elemental concentration of rocks and other geological formation. The analysis provides information about the geological formation of the region. Moreover, rock samples may be used as coarse aggregates in cement or concrete in the biological shield of a nuclear installation. NAA of the rock samples helps the researcher to estimate the long-lived activity that may be developed in the shield due to long-term operation. This study helps to plan the management of radioactive waste after decommissioning of the facility. NAA of rock samples from Rajasthan, India, carried out by Acharya et al. has helped to understand the formation of natural emeralds. NAA has important contribution in medicine: biochemically significant elements in trace amount can be effectively quantified using INAA. Variation in elemental composition induced by various diseases, particularly for malignant tissues, can be quantified using INAA. NAA with associated particle imaging (API) is a prospective tool for elemental analysis to monitor the growth of an animal in response to new genetic, pharmacologic procedure. The inherent high accuracy of NAA will offer it as one of the preferred techniques for elemental analysis at the ultra-trace level in human body fluid to follow the biochemical baseline values or the changes therein due to diseases.

Author details

Maitreyee Nandy

Address all correspondence to: mnandy07@gmail.com

Saha Institute of Nuclear Physics, Kolkata, India

References

[1] Pai BHV et al. Estimation of trace element concentration and neutron induced radioactivity in rock samples of different geological compositions for neutron shielding. Indian Journal of Pure & Applied Physics. 2016;**54**(01):7-14

[2] Kapadia AJ et al. Neutron stimulated emission computed tomography for diagnosis of breast cancer. IEEE Transactions on Nuclear Science. 2008;**55**(1):501-509

[3] Maglich BC, Nalcioglu O. 'ONCOSENSOR' for noninvasive high-specificity breast cancer diagnosis by carbogen-enhanced neutron femto-oximetry. ASME Conference Proceedings: Congress on NanoEngineering for Medicine and Biology. 2010; ISBN: 978-0-7918-4392-5. pp. 57-58. Copyright © 2010 by ASME

[4] El-Taher A. INAA and DNAA for uranium determination in geological samples from Egypt. Applied Radiation and Isotopes. 2010;**68**(6):1189-1192. DOI: 10.1016/j.apradiso.2010.01.046

[5] Uwah EJ, Rosenberg RJ. Measurement of radioelement contents of rocks of Ugep, S.E. Nigeria, by gamma-ray spectrometric and instrumental neutron activation analysis techniques. Applied Radiation and Isotopes. 1993;**44**(5):855-858. DOI: 10.1016/0969-8043(93)90028-9

[6] Medhat ME, Fayez-Hassan M. Elemental analysis of cement used for radiation shielding by instrumental neutron activation analysis. Nuclear Engineering and Design. 2011; **241**(6):2138-2142. DOI: 10.1016/j.nucengdes.2011.03.025

[7] Nazarov VM, Frontasyeva MV, Lavdanskij PA, Stephanov NI. NAA for optimization of radiation shielding of nuclear power plants. Journal of Radioanalytical and Nuclear Chemistry. 1994;**180**(1):83-95. DOI: 10.1007/BF02039906

[8] Alden JR, Minc L, Lynch TF. Identifying the sources of Inka period ceramics from northern Chile: Results of a neutron activation study. Journal of Archaeological Science. 2006;**33**(4): 575-594. DOI: 10.1016/j.jas.2005.09.015

[9] Huckell BB, Kilby JD, Boulanger MT, Glascock MD. Sentinel Butte: Neutron activation analysis of White River Group chert from a primary source and artifacts from a Clovis cache in North Dakota, USA. Journal of Archaeological Science. 2011;**38**(5):965-976. DOI: 10.1016/j.jas.2010.11.011

[10] Michelsen OB, & Steinnes E. Determination of copper in geological material by neutron activation and gamma-gamma coincidence spectrometry. Talanta. 1968;**15**(6):574-578. DOI: 10.1016/0039-9140(68)80135-3

[11] Ravisankar R, Eswaran P, Seshaderssan NP, Rao B. Instrumental neutron activation analysis of beachrock samples of South East Coast of Tamilnadu, India. Nuclear Science and Techniques. 2007;**18**(4):204-211. DOI: 10.1016/S1001-8042(07)60047-5

[12] Ravisankar R, Manikandan E, Dheenathayalu M, Rao B, Seshadreesan NP, Nair KGM. Determination and distribution of rare earth elements in beach rock samples using instrumental neutron activation analysis (INAA). Nuclear instruments and methods in physics research section B: Beam interactions with materials and atoms. Nuclear Instruments and Methods in Physics Research B. 2006;**251**(2):496-500. DOI: 10.1016/j.nimb.2006.07.021

[13] Acharya RN, Mondal RK, Burte PP, Nair AGC, Reddy NBY, Reddy LK, Manohar SB. Multi-element analysis of emeralds and associated rocks by ko neutron activation analysis. Applied Radiation and Isotopes. 2000;**53**(6):981-986. DOI: 10.1016/S0969-8043(99)00272-9

[14] El-Taher A, Alharbi A. Elemental analysis of natural quartz from Um Higlig, Red Sea Aea, Egypt by instrumental neutron activation analysis. Applied Radiation and Isotopes. 2013; **82**:67-71. DOI: 10.1016/j.apradiso.2013.07.002

[15] El-Taher A. Elemental analysis of granite by instrumental neutron activation analysis (INAA) and X-ray fluorescence analysis (XRF). Applied Radiation and Isotopes. 2012; **70**(1):350-354. DOI: 10.1016/j.apradiso.2011.09.008

[16] El-Taher A. Rare-earth elements in Egyptian granite by instrumental neutron activation analysis. Applied Radiation and Isotopes. 2007;**65**(4):458-464. DOI: 10.1016/j.apradiso.2006.07.014

[17] El-Taher A. Determination of chromium and trace elements in El-Rubshi chromite from Eastern Desert, Egypt by neutron activation analysis. Applied Radiation and Isotopes. 2010;**68**(9):1864-1868. DOI: 10.1016/j.apradiso.2010.04.018

[18] El-Taher A, Kratz KL, Nossair A, Azzam AH. Determination of gold in two Egyptian gold ores using instrumental neutron activation analysis. Radiation Physics and Chemistry. 2003;**68**(5):751-755. DOI: 10.1016/S0969-806X(03)00401-8

[19] Schwedt A, Aravantinos V, Harami A, Kilikoglou V, Kylafi M, Mommsen H, Zacharias N. Neutron activation analysis of Hellenistic pottery from Boeotia, Greece. Journal of Archaeological Science. 2006;**33**(8):1065-1074. DOI: 10.1016/j.jas.2005.11.009

[20] Srivastava A, Bains GS, Acharya R, Reddy AVR. Study of seleniferous soils using instrumental neutron activation analysis. Applied Radiation and Isotopes. 2011;**69**(5):818-821. DOI: 10.1016/j.apradiso.2011.01.027

[21] Steinhauser G, Sterba JH, Bichler M, Huber H. Neutron activation analysis of Mediterranean volcanic rocks–an analytical database for archaeological stratigraphy. Applied Geochemistry. 2006;**21**(8):1362-1375. DOI: 10.1016/j.apgeochem.2006.03.012

[22] Wasserman JC, Figueiredo AMG, Pellegatti F, Silva-Filho EV. Elemental composition of sediment cores from a mangrove environment using neutron activation analysis. Journal of Geochemical Exploration. 2001;**72**(2):129-146. DOI: 10.1016/S0375-6742(01)00158-3

[23] Watterson JIW, Sellschop JPF, Erasmus CS, Hart RJ. The combination of multi-element neutron activation analysis and multivariate statistics for characterisation in geochemistry. The International Journal of Applied Radiation and Isotopes. 1983;**34**(1):407-416. DOI: 10.1016/0020-708X(83)90140-0

[24] Zhu J, Glascock MD, Wang C, Zhao X, Lu W. A study of limestone from the Longmen Grottoes of Henan province, China by neutron activation analysis. Journal of Archaeological Science. 2012;**39**(7):2568-2573. DOI: 10.1016/j.jas.2012.03.010

[25] Contis ET. Analysis of drinking water near and far from thermal springs using instrumental neutron activation analysis. Developments in Food Science. 1995;**37**:2109-2127. DOI: 10.1016/S0167-4501(06)80277-3

[26] Arriola H, Longoria L, Quintero A, Guzman D. INAA of trace elements in colorectal cancer patients. Biological Trace Element Research. 1999;**71**(Issue 1):563-568

[27] Ebrahim AM Study of Selected Trace Elements in Cancerous and Non-Cancerous Human Breast Tissues Using Neutron Activation Analysis [Thesis]. University of Khartoum; 2003 http://www.iaea.org/inis/collection/NCLCollectionStore/_Public/38/059 /38059170.pdf

[28] Rees JR, Scot Zens M, Gui J, Celaya MO, Riddle BL, Karagas MR. Non Melanoma Skin Cancer and Subsequent Cancer Risk. PLoS One. 2014;**9**(6):e99674. DOI: 10.1371/journal.pone.0099674

[29] Zaichick V, Zaichick S. Trace Element Levels in Prostate Gland as Carcinoma's Markers. Journal of Cancer Therapy. 2017;**8**(2). DOI: 10.4236/jct.2017.82011

[30] Zaichick V Ye, Tsyb AF, Vtyurin BM. Trace elements and thyroid cancer. Analyst. 1995; **120**(3):817-821

[31] Zaichick V, Zaichick S. Age-Related Changes of Trace Element Contents in Intact Thyroid of Females Investi- gated by Neutron Activation Analysis. HSOA Journal of Gerontology and Geriatric Medicine. 2017;**3**:15. DOI: 10.24966/GGM-8662/100015

[32] Aloia JF, Ma R, Vaswani A, Feuerman M. Total-Body Calcium Estimated by Delayed Gamma Neutron Activation Analysis and Dual-Energy X-ray Absorptiometry. Osteoporosis International. 1999;**10**(6):510-515

[33] Davis SS, Hardy JG, Newman SP, Wilding IR. Gamma scintigraphy in the evaluation of pharmaceutical dosage forms. European Journal of Nuclear Medicine. 1992;**19**(Issue 11): 971-986

[34] Viana RS, Valverde MG, Mekkaoui C, Yoriyaz H, Jackowski M. NSECT sinogram sampling optimization by normalized mutual information. Progress in Biomedical Optics and Imaging - Proceedings of SPIE. 2015;9412. DOI: 10.1117/12.2082496

Chapter 3

Colombian Neutron Activation Analysis Laboratory (CNAAL): Applications and Development Using the Nuclear Research Reactor IAN-R1

Guillermo Parrado, David Alonso, Julián Orozco,
Mary Peña, Andrés Porras, Martha Guzmán,
Nelson Acero and Mauricio López

Additional information is available at the end of the chapter

http://dx.doi.org/10.5772/intechopen.74395

Abstract

This chapter describes recent advances of the Colombian Neutron Activation Analysis Laboratory (CNAAL) within the framework of expansion of geoscientific and nuclear knowledge in Colombia. Having the necessary historical references as a pillar of the current developments, the authors initially describe technical facilities of the laboratory and then articulate in an integral way of the value chain of this singular scientific and technological installation of Colombia. Its different stages beginning with the preparation of the samples, its irradiation process by different systems, recent novelty of the development of an "automated system of positioning of samples for gamma spectrometry," analysis of gamma spectra to obtain concentration data of chemical elements, management of the radioactive waste generated, analytical quality control of the data obtained and finalizing in the use of this data to cover selected topics of knowledge in strategic sectors of our country's development like a sustainable exploitation of mineral and hydrocarbons resources, researches in environmental and forensic sciences, technical developments in nuclear sciences, all aimed at improving the quality of life of Colombian citizens.

Keywords: neutron activation analysis, nuclear research reactor, applications, method validation, rare earth elements

1. Introduction

The Colombian Neutron Activation Analysis Laboratory (CNAAL) is a facility used for qualitative and quantitative nondestructive chemical multielemental analysis by activating samples at the Nuclear Research Reactor IAN-R1 and analyzing their decay products using Gamma Spectrometry. Neutron activation analysis (NAA) in Colombia started at the Institute of Nuclear Affairs (IAN) when the nuclear reactor research IAN-R1 first achieved criticality in 1965. This technique has been used for over 30 years mainly for elemental analysis with applications in geology, hydrology, environmental and forensic sciences. In 1998, the Colombian government closed down the Reactor and the country's nuclear development fell behind other Latin American countries who also began their nuclear research applications in the 1960s.

The onset of NAA in Colombia began in the mid-1950s, not long after the United States President Dwight D. Eisenhower gave his atoms for peace speech at the United Nations General Assembly on December 8, 1953. The Atoms for peace program served as shorthand for a number of programs intended to spread the peaceful uses of nuclear physics around the world and demonstrating its usefulness in the fields of medicine and energy generation. In 1955, the Colombian Institute of Nuclear Affairs (ICAN) was created, but it was not until 1957 that the Radioactive Analysis Laboratory was built as part of the ICAN chemistry program [1], being the predecessor of the current Neutron Activation Analysis Laboratory. This laboratory had basic instrumentation for personnel training purposes and was used for radiometric analysis by low-resolution gamma spectrometry using scintillation detectors like NaI(Tl) and single-channel systems.

Colombia was admitted to the International Atomic Energy Agency (IAEA) in the year 1960 and in 1964 began the construction of the dome building where the nuclear reactor stands today, later that same year the country received nuclear fuel manufactured in the United States as a donation by the American government. On January 20, 1965, the nuclear reactor achieved its first criticality (**Figure 1**) and operated at 10 kW during its first months [2].

In 1965, the first NAA samples were irradiated for elemental analysis [3], and the results were published 2 years later in 1967. Many scientific papers were published during this time, out of which two were of particular interest: the determination of inorganic iodine in samples of urine and the determination of trace amounts of selenium and tellurium in sulfur samples [4, 5]. The year 1971 witnessed the arrival of the first high-resolution Gamma Spectrometry System consisting of a germanium-lithium detector coupled to a multichannel analyzer (**Figure 2**). During this year, research focused on petrochemical analysis with the determination of vanadium in oil samples followed by the irradiation of food samples to determine nitrogen content in cereals.

Due to the boom in radioactive mineral exploration activities, the then Institute of Nuclear Affairs promoted a program for the quantitative evaluation of uranium and thorium in Colombia, for which Neutron Activation Analysis was used for the study of radioactive minerals and elemental determination in geological samples [1, 6]. Given the increased demand for this type of analysis and the interest from different companies on radioactive mineral exploration

Figure 1. Nuclear reactor start-up by Colombian president Guillermo León Valencia (1965).

in the Colombian territory, a similar nuclear technique was implemented at the time which helped improve the precision of analysis: delayed neutron counting (DNC) (**Figure 3**). Delayed neutrons are emitted after nuclear fission events by one of the fission products sometime after the fission process [7].

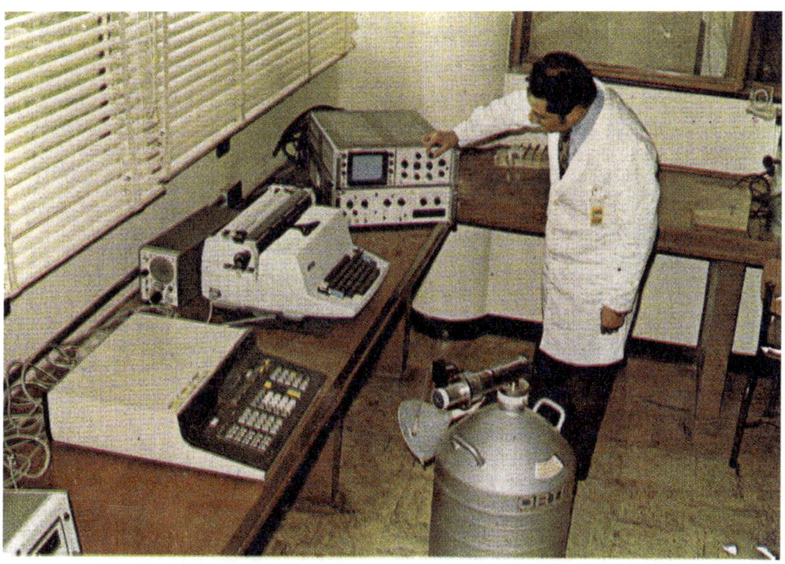

Figure 2. Neutron activation analysis instrumentation, Ge-li detection system (1971).

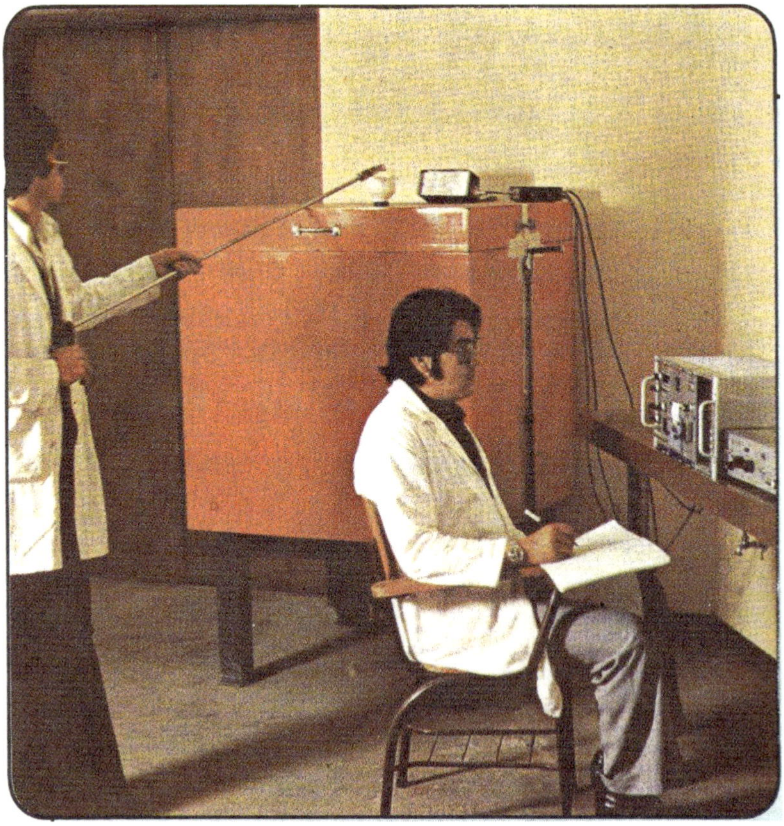

Figure 3. U/Th analysis by DNC (1974).

As of 1973 and thanks to the support offered by the French Government, NAA was implemented as an alternative to conventional analytical methods for the determination of small quantities of precious metals such as gold, silver, platinum and palladium, among others [8]. For 12 years, from 1975 to 1986, the NAA technique reached its stage of the greatest development and was used as standard for elemental analysis. Thanks to the support given by the IAEA, the country received funds to upgrade its NAA laboratory due to the potential growth of users demanding this type of analysis (**Figure 4**). During this period, the use of NAA in Forensic Sciences was also introduced.

From 1986 to 1989, work focused on improving procedures and methodologies in the application of NAA for mineral resource exploration and studies of sediments and water pollution [9–11]. From 1987 to 1990, there was a drastic decrease in workload (**Figure 5**) due to an upgrade at the nuclear reactor.

Once the reactor was up and running, the laboratory continued its routine analysis, providing services to internal projects as well as to external clients. NAA was used mainly for mining companies and special forensic studies (**Figure 6**) [9]. In 1992, the laboratory was moved to a new space, which was built as an annex to the Reactor's Building with the sole purpose of installing a pneumatic transfer system that would allow for the measurement of short-lived radionuclides (with average half-lives of the order of minutes and seconds).

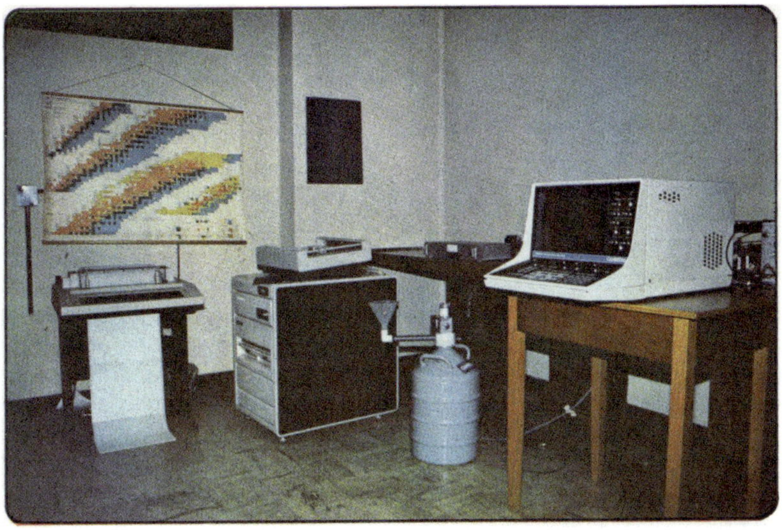

Figure 4. NAA instrumentation (1976).

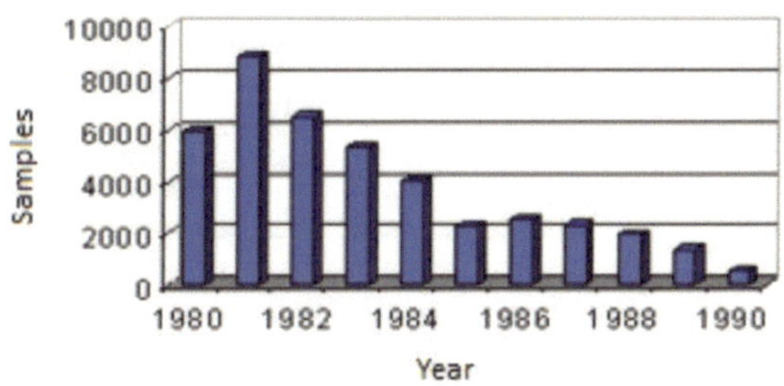

Figure 5. Irradiated samples from 1980 to 1990 [12].

Figure 6. NAA laboratory (1989).

In September 1994, the nuclear reactor went into an extended shutdown period due to modernization of its instrumentation and control systems as well as conversion from highly-enriched uranium (HEU) fuel to low-enriched uranium (LEU) TRIGA fuel. Process took place from 1995 to 1997.

In spite of the liquidation in 1997 of the Institute of Nuclear Sciences and Alternative Energies (Formerly Institute of Nuclear Affairs), the nuclear reactor still operated with its new TRIGA fuel and was utilized for the analysis of forensic samples needed by the Police Department for the determination of barium, antimony and copper. During the first quarter of 1998, forensic samples were analyzed for the determination of mercury in hair, but due to the closure of the institute, the Reactor was shut down on March 31 that same year [3]. This series of events halted the operational experience that the NAA Laboratory had built for about 32 years (**Figure 7**).

It was not until 2005 that the Colombian Government decided to restart its Nuclear Program and began training personnel at the Reactor and associated nuclear laboratories. In May 2006 with support from the IAEA, the NAA technique was finally resumed, and tests were performed for the two Gamma Spectrometry Systems available at the time. One system equipped with a NaI(Tl) scintillation detector and the other with a Canberra 7229P HPGe semiconductor detector.

However, the NAA laboratory formally resumed activities in 2009, when the authorization for radioactive material handling was granted by the National Regulatory Authority.

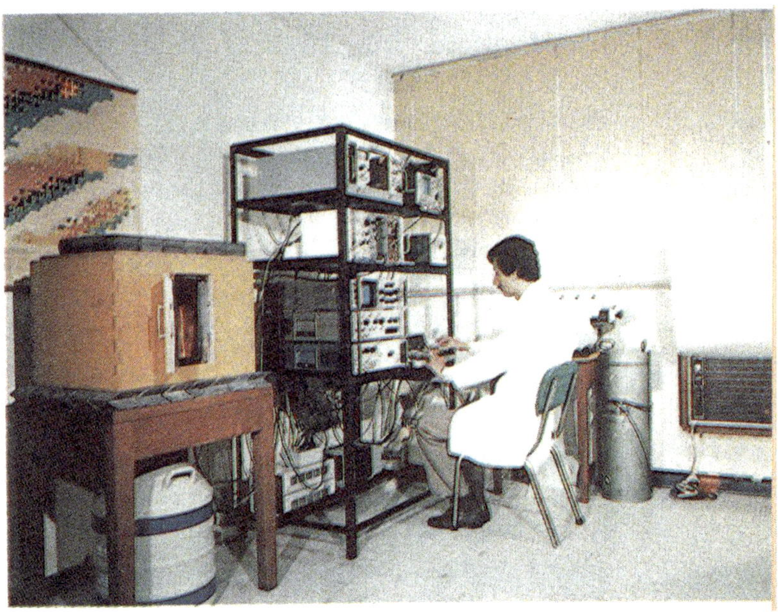

Figure 7. Gamma spectrometry system (1997).

Several expert missions were received for training of new personnel. The first objective of the Laboratory in this new stage was to provide the service of multielement analysis of geological samples, for which the relative method (comparator method in the literature) is used through certified reference materials (CRM) [13, 14]. **Figure 8** shows the state of the laboratory in 2009.

During the years of 2013 and 2015, the gamma spectrometry systems were modernized with High-Purity Germanium (HPGe) Canberra® detectors (**Figure 9**), an automated positioning system was also installed in the laboratory, making it unique in Latin America. IAEA experts have been continuously assisting on the validation for the test method and neutron flux characterization of the core. Today, thanks to the support given by the IAEA, the nuclear reactor and NAA laboratory have some of the most modern installations in Latin America with strong future prospectives in various fields of science.

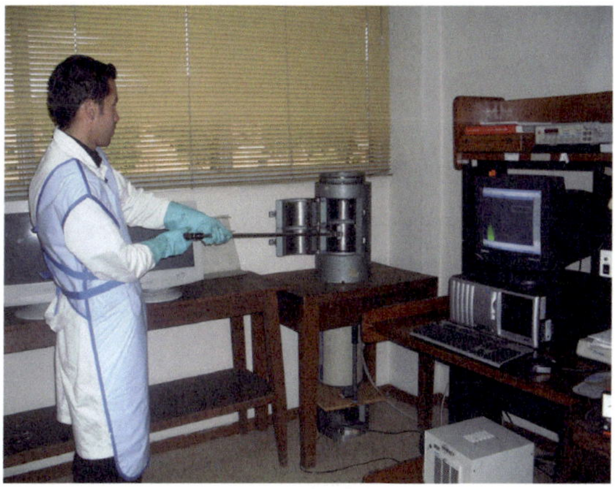

Figure 8. NAA laboratory (2009).

Figure 9. NAA laboratory (2017).

2. Laboratory description

Neutron activation analysis is a multielemental chemical analytical technique based on neutrons generated by the nuclear reactor to create radioactive isotopes from stable isotopes in a sample material. The technique relies on excitation by neutrons so that the treated sample emits gamma-rays, and this radiation is then analyzed enabling the user to detect, identify and measure the presence of radioactivity in natural or man-made sources. The main use for the laboratory is to complement the conventional analytical techniques adopted in the institute, especially for those elements whose routine determination may require costly procedures with high environmental impact due to their nature and complexity. This type of analysis is used for the elaboration of a national geochemical map, which is essential for mineral exploration in the territory.

In 2009, not long after restarting the IAN-R1 Research Reactor, the laboratory was re-established at the Colombian Geological Survey, serving the country once again as a key player in the determination of elemental composition in geological matrices. Samples are irradiated under appropriate safety conditions following national regulations, which are lined up to the International Atomic Energy Agency (IAEA) and International Commission on Radiation Protection (ICRP) guides and scientific publications.

From 2016 to 2017, the delayed neutron counting technique was re-established for the determination of uranium and thorium in resource exploration projects due to the sensitivity of the technique, which is lower than 1 μg/kg and can be used to analyze materials with high uranium content (including U_3O_8) and enrichment of ^{235}U [15].

The scientific staff are qualified and trained with several years of experience and extensive operational knowledge in the management of radioactive material, radiation protection, qualitative and quantitative chemical analysis, waste management, isotope applications and nuclear energy applications.

A brief description of the rooms that make up the CNAAL is given in the following sections.

2.1. Sample preparation room

This room has the necessary infrastructure and equipment for the preparation and adaptation of samples. A Niton XL3t GOLDD portable X-ray fluorescence analyzer, which is used for the preliminary characterization of the samples, a homogenizer mill, three analytical balances, a tablet press and a bag sealer. Reference materials and standards are properly stored under controlled conditions in this room.

2.2. Neutron activation analysis room

This room is where the samples are sent into and out of the core if the pneumatic transfer system is to be used. The systems console and Port No. 1 are located here. There is a gas and vapor extraction cabin with a 1.5 cm thick shielded port used to receive irradiated material from the reactor, and there is also a leaded glass which protects the staff from radiation exposure from the samples (**Figures 10** and **11**). Verification sources used for radiation monitors and calibration of gamma spectrometers are also stored in this room.

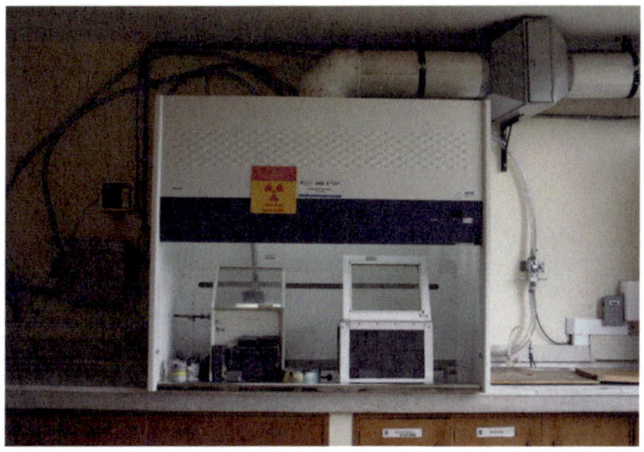

Figure 10. Gas extraction cabin.

2.3. Gamma spectrometry rooms 1 and 2

The Gamma Spectrometry Systems used in the detection and quantification of the gamma-rays emitted by the activated samples after irradiation are located in two different rooms. A HPGe Canberra GC-1020 detector, a Canberra 2002CSL pre-amplifier and the InSpector 2000 multichannel module are located in the first room.

Figure 11. Shielded port for sample transfer.

Figure 12. Gamma spectrometry – Room 2.

The second room (**Figure 12**) has four HPGe detectors. Two Canberra GC-3018 detectors with 30% efficiency and energy resolution of 1.8 keV at 1.33 MeV at full-width at half-maximum (FWHM), and two GC-7020 units with 70% efficiency and energy resolution of 2.0 keV at 1.33 MeV (FWHM); each detector comes with its respective shielding, a LYNX® digital signal analyzer and is controlled by the Canberra's Genie2000 v.3.3 software. There is also an automated positioning system that uses a robotic arm to automatically place the samples in each of the four HPGe detectors and reads the gamma spectra during the time it is programmed; this system was designed for radiological protection purposes.

2.4. Delayed neutron counting room

he room assigned for delayed neutron counting consists of a console and Port No. 2 of the pneumatic transfer system, one ton of paraffin shield which sits on top of 20 cm thick concrete blocks, a geometric arrangement of eight proportional BF_3 counters and its associated electronic instrumentation for neutron counting and determination of uranium and thorium in geological, environmental and forensic matrices. There is a central hole in the paraffin shield were samples are placed for reading, and there is also a manual mechanism for sample extraction once readings are done (**Figure 13**).

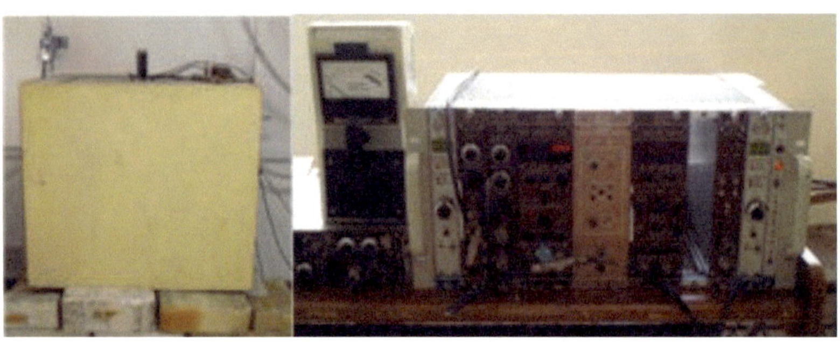

Figure 13. Delayed neutron instrumentation.

2.5. Decay room

The radioactive material decay room is a space with lead and concrete shields needed to store activated samples for decay. It has two cylindrical lead containers with 6 cm thick walls for the storage of radioactive waste and two compartments made out of 15 cm thick concrete blocks for the storage of samples according to their half-life (**Figure 14**). Activated samples are temporarily stored in this room until the exemption levels are reached [16].

2.6. Pneumatic transfer system

The pneumatic transfer system allows for the rapid exchange of samples between the neutron activation room, the delayed neutron counting room and the nuclear reactor. Its master control is located in the Reactor's console room and without authorization from the reactor's personnel, it is not possible to send samples for irradiation, and this results in a redundancy in the safety of the sample positioning system.

This system consists of two compressors: flow diverters, two controls to send and receive samples (**Figure 15**) and a high-density polyethylene duct. The samples enter directly into one of the two aluminum terminals located in the core (positions D3 and C4), where the highest neutron flux can be found. It has the advantage to transfer activated samples to different areas speeds of 15 m/s, being an important aspect in the radiological protection of the personnel. This system is equipped with "air cushion" braking mechanisms to avoid violent crashes against the system ports.

The systems control unit includes a digital counter which can be set up for times between 30 s and 4 h. This unit controls the automatic valves to open and cut the air pressure at the right time. Since it is a complex pneumatic system with two receiving stations and two in-core terminal positions, there are diverters operated remotely from the control unit, which allows for the selection of irradiation positions and terminal stations where samples are received.

Due to the production of Ar41 during activation, terminal ports in the laboratory rooms are located inside extraction cabins with filters for radionuclides, preventing the inhalation of the radioactive gas by the operators.

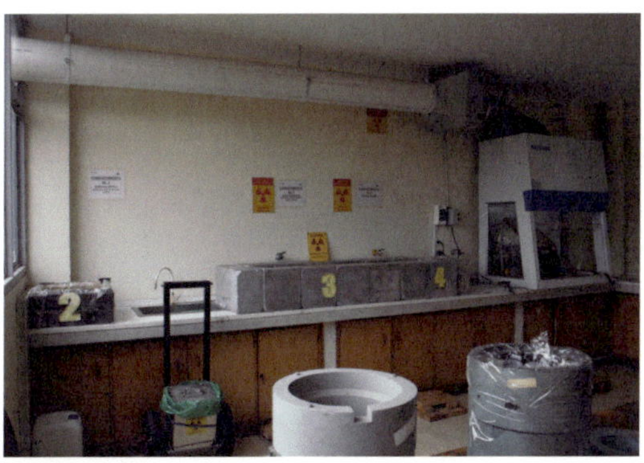

Figure 14. Concrete shielding in decay room.

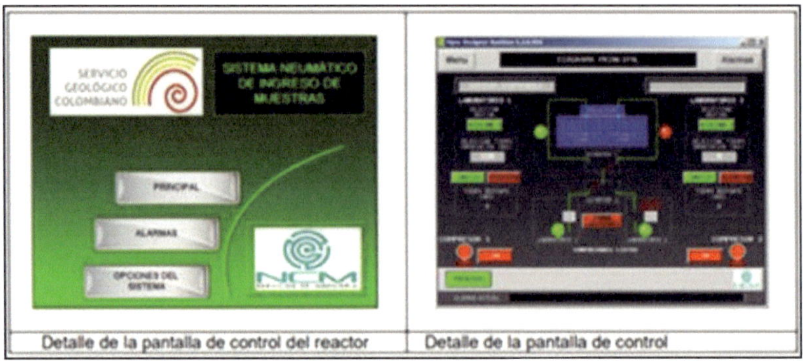

Figure 15. Pneumatic transfer system controls.

3. Sample preparation/irradiation

The Colombian Geological Survey serves the country by providing reliable scientific data through research into basic and applied subsoil geo-sciences; evaluating and monitoring threats of geological origin; exploring and monitoring petroleum resources, minerals and groundwater; the ability to study the elemental composition of samples such as rocks, soils, sediments, minerals, water and gases is a major asset. Collected samples are often taken to different laboratories for a variety of analyses if required by the research being conducted. Several analytical techniques ranging from Gravimetric Analysis and Atomic Absorption Spectrometry, to X-ray Fluorescence, Inductively Coupled Plasma Mass Spectrometry (ICP-MS), and Neutron Activation Analysis among others are available at the Colombian Geological Survey.

Geological materials analyzed by NAA need to be previously dried at room temperature, crushed, pulverized and sieved to a particle size of 150 μm (100 mesh, **Figure 16**). Once the sample is received (~ 50 g), moisture content needs to be determined in order to make future corrections referencing the dried sample. Samples are homogenized, weighed (0.250 ± 0.001 g), pressed and encapsulated in plastic hermetically sealed polyethylene vials.

Figure 16. Sample preparation process: (a) drying (b) grinding and (c) screening.

http://dx.doi.org/10.5772/intechopen.74395

Samples for long-lived element activation (days to years) are placed at the periphery of the core and vials with samples are arranged in racks as shown in the following diagram (**Figure 17**). These racks are placed in vacuum-sealed Ziploc bags before irradiation in the G3-G4 positions (**Figure 18**).

The following elements can be determined after a 4-h irradiation operating at 30 kW: Sm, Lu, U, La, Nd, Eu, Hf, Ce, Yb, As, Sb, Ba, Br, Cd, Gd, Ga, Ho, Mo, W, Th, Cr, Cs, Sc, Ir, Ni, Se, Ag, Ta, Tb, Tm, Rb, Fe, Co, Zn, Zr. The neutron flux is measured by 5 mg Al + 0.1% Au rectangular foils as previously shown in **Figure 17**. Measurement required to obtain the correction factor fϕ. Samples for short-lived element activation (seconds to a few hours) are irradiated inside the core at positions D3 or C4 (**Figure 18**). These samples are encapsulated in cylindrical pressure-sealed polyethylene containers, packed in pairs into rabbits (polyethylene vials) and transferred into the core by the pneumatic transfer system.

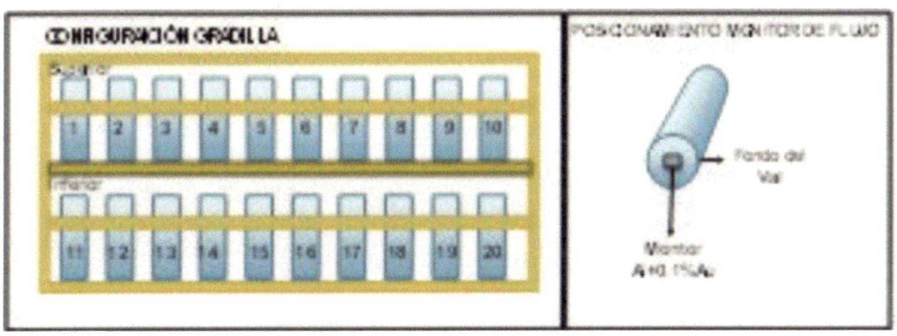

Figure 17. Rack sample configuration (left) neutron flux monitors attached to vials (right).

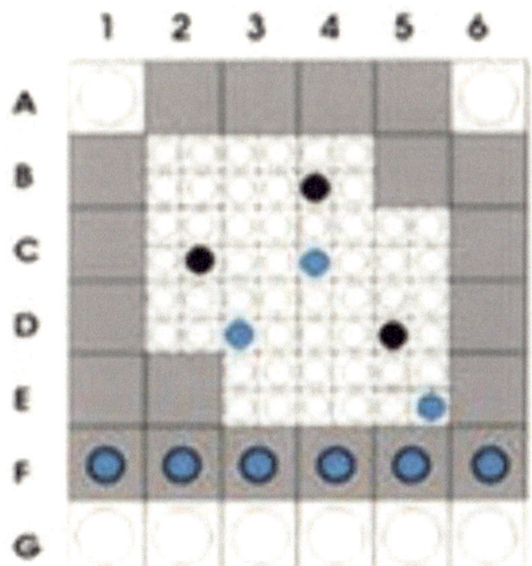

Figure 18. Reactor core schematic (IAN-R1).

Sample irradiation in these rabbits ranges from 40 s at 5 kW for the analysis of uranium and thorium using delayed neutron counting techniques and up to 5 min at 30 kW for the analysis of short-lived elements (Al, Ca, Mg, Ti, V, Dy, Mn, K and Na).

The IAN-R1 nuclear reactor was built by the Lockheed Western Export Company and was commissioned in 1965 as a graphite-reflected pool-type research reactor, cooled by natural convection with light water. The current core consists of 50 fuel elements made of U-$ZrH_{1.6}$ enriched up to 19.75%. The reactor is licensed by the Ministry of Mines and Energy (Nuclear Regulatory Body) to operate at the maximum steady-state power of 30 kW, and it is located inside a cylindrical tank made of carbon steel 6×10^{-3} m thick, 5.25 m tall and 2 m in diameter with capacity to store up to 16 m^3 of water.

The nuclear reactor's instrumentation and control systems were fully upgraded during 2012 and 2013 (**Figure 19**) by National Institute of Nuclear Research (ININ, México), in 2016 a new automated pneumatic transfer system (**Figure 20**) was installed, replacing the original system installed in 1997. This system opened up two irradiation positions inside the core, remotely

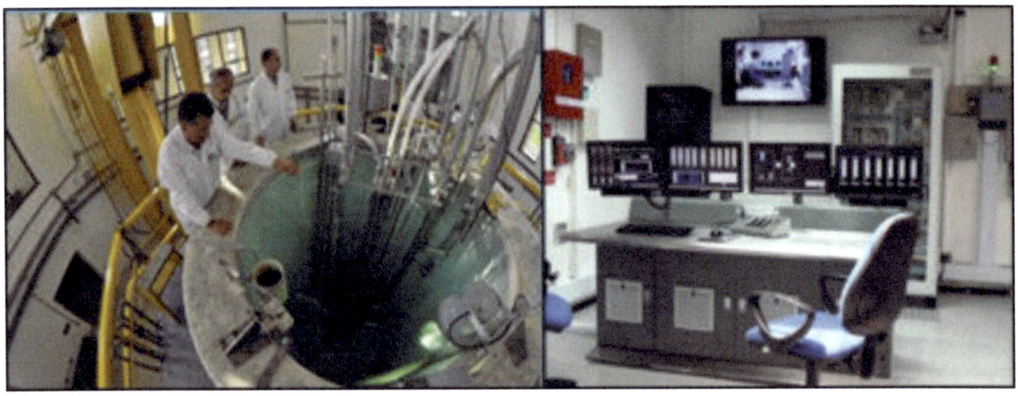

Figure 19. Nuclear research reactor IAN-R1.

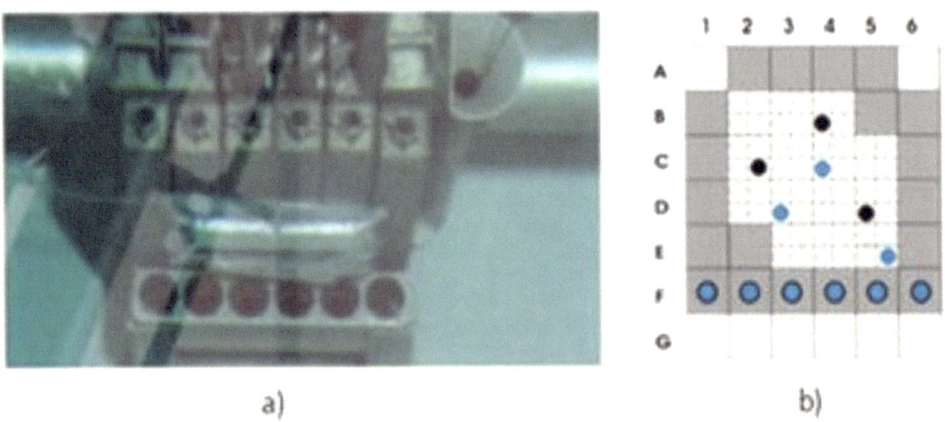

Figure 20. Rack with samples for irradiation.

controlled from ports 1 and 2 located in the neutron activation and delayed neutron counting rooms, respectively.

Additionally, there is a peripheral pneumatic transfer system that was part of the original design (1965) and is used to irradiate samples in a position adjacent to the core (Position A6). This old system was used from 1968 to 1992 for radioisotope production (^{24}Na, ^{32}P, ^{82}Br, ^{198}Au and ^{99}Mo) [17].

The rack containing the flux monitors and samples is positioned in the middle of the frontal face of the core, irradiated during 4 h at 30 kW (**Figure 20**) and subjected to a thermal neutron flux of around 2.3×10^{11} neutrons cm^{-2} s^{-1}.

Table 1 presents a summary of the experimental conditions used for the analysis of short-, medium- and long-lived elements.

Characteristic	Value/description	
	Pneumatic transfer system	**Rack system**
Method	Direct comparator	Direct comparator
Measurand	Mass fraction	Mass fraction
Sample/comparator (mass)	0.250 ± 0.001 g	0.200 ± 0.001 g
Reactor power	20–30 kW	30 kW
Irradiation position	D3 or C4	G3-G4
Irradiation time	1–5 min	4 h
Flux monitor	Al + 0.1% Au 0.5 mg	Al + 0.1% Au 0.5 mg
Decay time 1	5 min	3–7 days
Reading time 1	5 min	3 h
Reading 1 geometry	50 mm	30 mm
Decay time 2	60 min	21–28 days
Reading time 2	10 min	4 h
Reading 2 geometry	10 mm	15 mm
Photon energy (keV)	Nuclide dependent	Nuclide dependent

Table 1. Experimental irradiation conditions.

4. Automated sample positioning system for gamma spectrometry

Automation allows greater control of counting geometries, less error in positioning, increased productivity in the analyses, increased control and quality assurance of the analytical data and decreased doses received by the staff [18].

The automated system developed in the NAA laboratory fulfills the following objectives: Programming of analysis sequences, opening and closing of shields for sample positioning in the detector, and communication with the software for data acquisition. This has improved productivity by enabling 24/7 operation, and as a side benefit there is also less exposure to ionizing radiation.

This system is based on electromechanical components that can handle up to 64 sample readings in the same sequence of analysis, for which each step includes the collection of the sample from a sample rack, the positioning of the sample in the detector, data acquisition at pre-defined reading times, and the return of the sample to the rack where the other samples are located.

All of this is possible by means of a high precision positioning system based on linear actuators. The system is controlled by a human machine interface (HMI) where execution commands can be programmed. **Figures 21–24** show how everything is set up.

In principle, the presence of personnel in the gamma spectrometry room is limited while the positioning system is in operation. However, shielding for the sample rack is to be installed in the near future for radiological protection purposes.

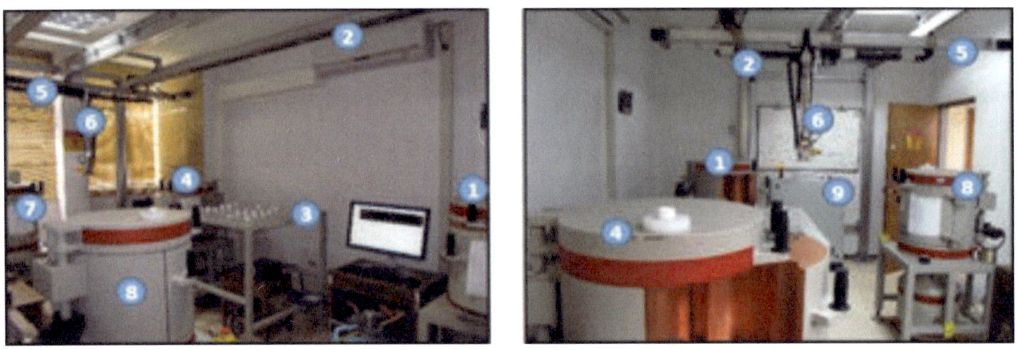

Figure 21. Positioning system: 1. Detector A; 2. X-axis linear actuator; 3. Sample rack (64 positions); 4. Detector B; 5. Y-axis linear actuator; 6. Z-axis linear actuator; 7. Detector C; 8. Detector D; 9. Control panel.

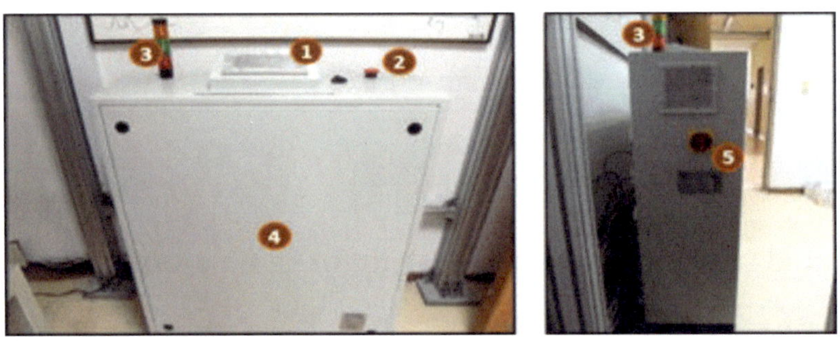

Figure 22. Positioning system: 1. HMI; 2. Emergency stop; 3. System status; 4. Electronic components; 5. Power control.

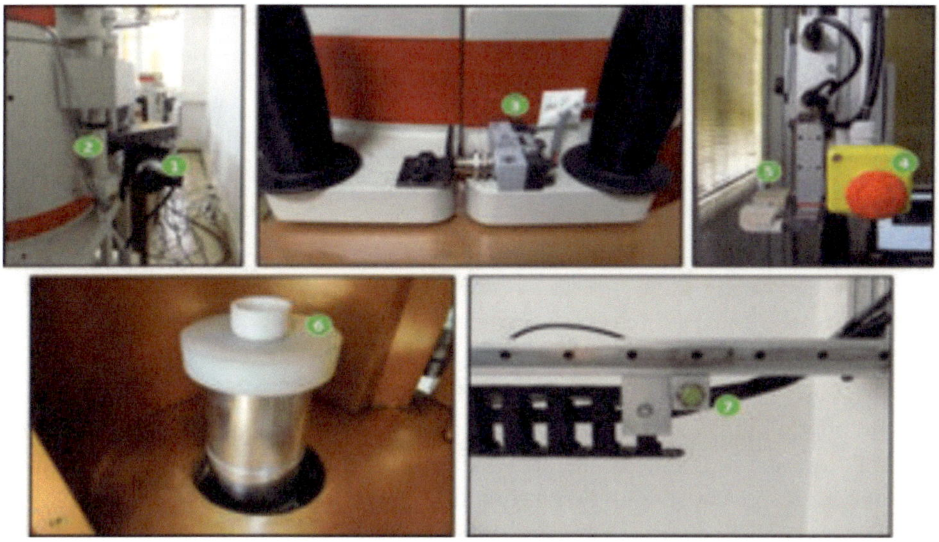

Figure 23. Positioning system: 1. Electric motor to open/close shielding; 2. Shield sensor for opening; 3. Shield sensor for closing; 4. Emergency stop; 5. Sample gripper; 6. Detector sample support; 7. Actuator motion limit sensor.

Figure 24. Positioning system.

Precision is determined by servomotors that provide the movement, which have a resolution of 1,048,576 pulses/revolution per axis. Coupled to the previously described servomotors, there are linear belt actuators (Accuracy ±1.0 mm) with their respective guides for alignment and friction reduction throughout the working area.

The system has two spatial adjustment options, point-to-point which displays the 64 positions of the sample rack and the 4 detector positions; and also single-point which is used to correct a common mismatch in all points using the point-to-point option, defining the first position of the rack, and allowing for the automatic adjustment of all the other positions. Both of these options are password protected for security reasons.

The point-to-point adjustment option must be used to change the counting geometry on the detectors, which must be performed prior to the execution of the sequences as required by the operators.

This positioning system greatly reduces manual efforts during the analysis of radioactive samples; the only manipulation required by our staff is the setup of the 64 samples in the rack. The idea of the use of this rack is minimizing Radiation Exposure, and thus enhancing the safety and well-being of personnel.

5. Data acquisition

In order to offer a quality service, the laboratory has made important updates in its instrumentation; acquiring four state-of-the-art HPGe Canberra Detectors for the measurement of gamma radiation and a novel sample positioning system.

The NAA relative method uses gamma radiation emitted by the radioactive nuclei from activated samples and compares it to the radiation emitted by a reference material with similar characteristics.

The characteristic gamma energies of each radionuclide are measured using one of the four solid-state semiconductor detectors (GC-3018 and GC-7020) coupled to LYNX® digital signal analysers controlled by the Canberra's Genie 2000 (v3.3) software. Each of these systems is calibrated weekly in energy, and its efficiency is checked monthly using a gamma check source kit consisting of ^{241}Am, ^{22}Na, ^{133}Ba, ^{137}Cs, ^{155}Eu and ^{60}Co electro-deposited point sources.

Depending on the irradiation, system used (**Table 1**), and once pre-determined decay times are reached, radiation measurement is performed by using one of the HPGe detectors [19].

Radiation measurement by samples coming from the pneumatic system is performed in some of the less efficient detectors (GC-1020 or GC-3018, depending upon availability). For complete analysis, two readings are carried out: the first, after 5 min of decay, positioning the vials individually at a distance of 50 mm from the detector and reading the corresponding spectra for 5 min. The second reading is done after the activity decays for an hour, at a distance of 10 mm and the spectra is read for 10 min. On the other hand, samples not going through the pneumatic transfer system are analyzed as follows: a first reading is done after a 4-day decay on one of the GC-3018 detectors, positioning the vials individually at a distance of 30 mm from the detector and reading the spectra for 3 h. A second reading is then performed after 21 days of decay on the higher efficiency GC-7020 detectors at a distance of 15 mm and the spectra is read for 4 h.

Flux monitors used during sample activation for flux corrections are also analyzed by gamma spectrometry on one of the GC-3018 detectors at a distance of 50 mm for 60 min. The samples obtained its corresponding gamma spectrum (**Figures 25** and **26**).

http://dx.doi.org/10.5772/intechopen.74395

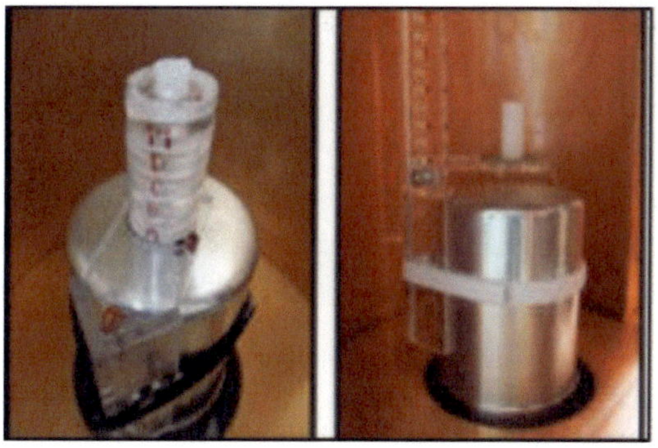

Figure 25. Detection geometry for check source verification.

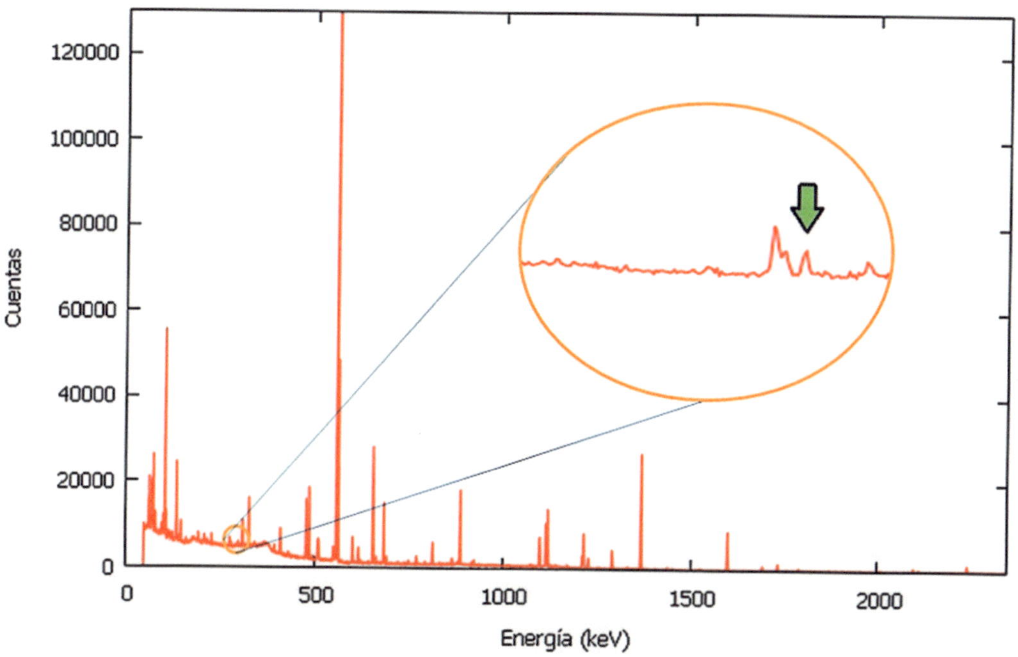

Figure 26. Gamma spectra from soil sample irradiated for 4 h at 30 kW and a 5-day decay.

The elemental determination in the NAA is done by using the relative calibration (direct comparator method) [20]. This method uses a sample with known mass of the elements of interest (comparator or standard) and an unknown sample which is irradiated simultaneously. Taking into account that the amount of radiation emitted from the activation of the sample is proportional to the neutron flux, and this in turn to the mass of the irradiated element, it is found that the ratio between the mass fraction of the element x and the amounts of influence is given by:

The subscripts indicate parameters for the unknown sample and the comparator or standard. That is, the mass fraction of the unknown sample of the element (measurand) and the mass fraction of the element in the reference material. W is the total mass of the samples. The number of net counts of the energy of interest (keV) and the counting time of the gamma radiation are decay correction factors of the peak; this factors are used to obtain the neutron flux correction factor, which quantify the gradient of the flux between the irradiation position of the sample and the comparator.

The neutron flux correction factor is determined as the ratio between the flux measured with the 0.1% Au-Al flux monitors at the sample position and that of the comparator. The neutron flux is proportional to the number of counts in energy range of 198Au and depends on other factors such as the neutron capture cross-section, isotope abundance, irradiation time, 198Au half-life, detection efficiency, and number of accounts for the emission energy of the radioactive isotope. Taking the relationship between the flux readings, all the terms except for the number of counts registered for the monitor at the sample position and for the monitor at the comparators position are canceled, obtaining:

This relationship is fulfilled, assuming that the irradiation conditions and counting geometries are similar.

6. Radioactive waste disposal

In compliance with national regulations, the Neutron Activation Analysis Laboratory has developed a simple scheme for the safe management of activated samples once they are no longer needed. These procedures are authorized and monitored by the National Nuclear Authority (Ministry of Mines and Energy). The aim of radioactive waste management is to isolate and apply protective measures to this type of waste so that there are no foreseeable future human health risks and/or negative effects on the environment.

Most of the radioactive waste corresponds to plastic materials (vials, racks, bags, tapes, etc.), as well as flux monitors and samples activated during irradiation, and also elements used in the decontamination of working areas (gloves, paper towels, plastic bags, etc.). Most of the waste is classified into three categories: Group 1 – Exempt Waste (EW), Group 2 – Very Short-Lived Waste (VSLW) and Group 3 – Very Low Level Waste (VLLW) [16].

The NAA Lab has a decay room for the temporary storage of VSLW, equipped with the necessary shielding and equipment for its safe handling. The room is locked and permanently monitored, admission is restricted to non-operating personnel, unless authorized otherwise. The stored waste is properly labeled and grouped into packages each workweek.

In order to classify and monitor temporarily stored waste, packages are analyzed by gamma spectrometry. Each package is analyzed separately in the GC-7020 detector at a reading geometry of 10 mm during 4 h, determining the activity of each of the radionuclides present.

Radioactive waste is discharged as conventional waste only when its activity reaches acceptable levels established by national regulations, with previous knowledge and consent of the Regulatory Authority. Procedures for the Management of Radioactive Waste are lined up with technical and administrative requirements established by national regulations. The following schematic allows for the safe management of radioactive samples and activated materials generated during practice.

a. Minimization: Only materials and samples directly linked to national projects are irradiated in order to generate useful information for the economic and social development of the country.

b. Segregation: Separating waste generated during short and long irradiations and that collected during decontamination activities.

c. Pre-treatment: Waste package preparation, 16–20 samples per package.

d. Classification: By activity concentration and half-lives of nuclides present in the sample.

e. Characterization: Classification and monitoring of temporarily store waste are performed by Gamma Spectrometric analysis of the packages generated each workweek. Every package is analyzed separately in Canberra HPGe GC-7020 detectors, at a reading geometry of 10 mm during 4 h, these reading are then analyzed and the activity of each radionuclide present is determined.

f. Storage: The NAA Lab has a decay room for the temporary storage of VSLW, equipped with the necessary shielding and equipment for its safe handling. The room is locked and permanently monitored, admission is restricted to non-operating personnel, unless authorized otherwise. The stored waste is properly labeled and grouped into packages each workweek.

7. Quality assurance

As part of the validation process for NAA using HPGe detectors and future accreditation of the laboratory under ISO/IEC 17025:2005 [21], the technique has been validated for the determination of rare earths such as La and Ce, and elements of interest such as U and Th in geological matrices. The following parameters were taken into account: selectivity, linearity, reproducibility, limits of detection and quantification, robustness and uncertainty estimation.

This process included the evaluation of detection limits and quantification of gamma radiation spectra obtained, according to the statistical criterion of Currie [22]. The following results show element concentrations in the sample in units of mg/kg. Ba: 129, Ce: 1.37, Co: 0.20, Cs: 0.29, La: 0.11, Rb: 10.06, Sb: 0.054, Sc: 0.024, Th: 0.27, U: 0.2. These results were comparable to those reported by other laboratories, thus demonstrating the competence of the NAA Laboratory on multielemental analysis in geological matrices.

Certified reference materials (rocks, soils, sediments and coals) are used as comparators for the implementation of the technique, and they serve as a basis to compare known quantities of an element with fractions of elements in the sample. Internal standards are used for routine control of the method, and these standards are geological materials with known concentrations reported by other laboratories worldwide that use NAA or other analytical techniques, these standards are evaluated under the same analytical conditions of the problem samples.

For the determination of uncertainty, the steps recommended in the Reference Guides [23, 24] were followed. First, the measurand was defined, establishing its relation with influence quantities and identifying them. Identifying those with greater contribution were evaluated according to their type: A or B. Finally, the combined uncertainty and the expanded uncertainties are quantified.

The uncertainty estimation was done following the bottom-up approach. The procedure consisted of establishing the measurand, identifying and quantifying the sources of uncertainty and finally determining the combined and extended uncertainties. The sample's mass, standard's mass, neutron flux gradient, counting geometry differences, and sample count statistics were evaluated as sources of uncertainty. Sample and standard count statistics as well as differences in irradiation geometry were identified as the main contributors to the uncertainty [25, 26]. The combined relative uncertainty for the studied elements oscillates between 2 and 8% (**Table 2**).

The results of the evaluation of performance: limits of detection, intermediate precision, robustness, veracity and uncertainty, meet the requirements established for the test method;

Element	Isotope	Half-life (days)	Energy 1 (keV)	Energy 2 (keV)	LOD (mg/kg)	Precision (%)	Veracity (%)	Uncertainty (%)
La	La-140	1.68	1596.2	487.0	0.11	6.1	97.4	4.0
Sb	Sb-122	2.72	692.8	–	0.05	9.5	103	1.9
U	Np-239	2.36	106.1	228.2	0.25	6.9	100	8.9
Ba	Ba-131	11.5	216.1	496.3	129	5.1	96.1	8.8
Ce	Ce-141	32.5	145.4	–	1.37	2.8	101	9.9
Co	Co-60	1925	1173.2	1332.5	0.20	3.2	95.6	5.8
Cs	Cs-134	754	604.7	795.9	0.29	2.1	103	6.9
Rb	Rb-86	18.6	1076.6	–	10.1	5.8	95.9	6.7
Sc	Sc-46	83.8	889.3	1120.5	0.02	3.5	98.3	6.2
Th	Pa-233	27.0	312.2	–	0.27	3.3	97.6	8.3

Table 2. Multi-elemental validation results.

establishing a line of work toward further validation of more elements and offering the scientific community a proven method according to international standards [27].

As part of the application for the accreditation process under the ISO/IEC-17043 standard [29], and the need for constant validation in performance and quality assurance, the laboratory participates in annual IAEA-WEPAL (Wageningen Evaluating Programmes for Analytical Laboratories) Proficiency Tests and Inter-laboratory Comparisons, obtaining excellent results and positioning its metrological competence, this being a major step toward accreditation under ISO/IEC 17025 [21].

8. Data applications

The Colombian Laboratory for Neutron Activation Analysis, CNAAL, is an installation oriented to the generation of high-quality analytical data that contribute to the geoscientific knowledge of the national territory, represented in the characterization of our valuable mineral and hydrocarbon resources. This potential of CNAAL's analytical technique can now be applied in vast areas of the country, which for decades were the scene of a long, costly and painful armed conflict, which ended in 2016 with the signing of the Peace Agreements between the Colombian State and the FARC guerrillas, the oldest in our continent.

Our laboratory has focused its analytical capabilities on the exploration of rare earth elements, which according to the OECD study [30] present a relatively favorable scenario for the search for these strategic minerals that present a greater supply risk taking into account its typical scarcity.

Rare earth elements (REEs) are central in information and communications technologies and green technologies, which is one of many reasons that justify studies in this area. In this way also, OECD's Cost of Inaction and Resource Scarcity; Consequences for Long-term Economic Growth (CIRCLE) Project "...aims at identifying how feedback from poor environmental quality, climate change and natural resource scarcity are likely to affect economic growth in the coming decades" [28].

Additionally, the characteristic mobility of REE is useful for the study of petrogenetic processes and the study of the geochemical cycle of uranium and other associated energy minerals.

Other applications planned for neutron activation analysis technique are related to: advances in the validation of analytical methods to determine elements, quality assurance by ISO/IEC 17025, continue with successful participation in the IAEA – WEPAL proficiency test and promotes future developments to generate impact researches on selected topics on geological materials characterization (rocks, soils, sediments, minerals and hydrocarbons), forensic sciences (element traces in crime scenes), archeometry (studies of provenance of bones, paintings, pottery, coins) and environmental sciences (mobility and accumulation of eco-toxic elements in humans, from technological and industrial processes and evaluation of environmental impacts in the biotic components), among others. In order to consolidate the credibility of our results, the LAAN has participated in a series of intercomparison exercises whose results have been improved to date.

Recent advances in the characterization of the neutron flux (thermal and epithermal) of the modern Colombian nuclear reactor IAN-R1 (upgraded in 2015), allow the CNAAL a window of opportunity for the implementation of the "k0 – NAA method", to improve the analytical capabilities of the laboratory, placing it at the level of other facilities of similar characteristics in other countries [29].

9. Conclusions

The Colombian Neutron Activation Analysis Laboratory and the Nuclear Research Reactor IAN-R1 area singular scientific and technological facilities in our country, located at the Colombian Geological Survey (Servicio Geológico Colombiano). CNAAL plays an exceptional role oriented to cover technical topics in sustainable exploitation of mineral resources and hydrocarbons, developments in nuclear sciences and researches in environmental and forensic sciences. At this point, the laboratory is available to the entire scientific and academic community of the country and recently upgrading of the Nuclear Research Reactor IAN-R1 (2014), also contributes to these national goals.

The CNAAL has state-of-the-art technology and competent personnel, which allows it to expand the coverage of research services in the country, through the Neutron Activation analysis in the execution and development of research projects.

Acknowledgements

Special mention is offered to the International Atomic Energy Agency – IAEA for its permanent support to CNAAL, represented in technical assistance, expert missions, equipments and spare parts, trainings and scientific visits in order to strengthen our scientific capacities. Thanks are due to Mr. Peter Bode (TU-Delft) for his invaluable assistance. Acknowledgement is made to Mr. Oscar Paredes Zapata, General Director of Colombian Geological Survey, for his assistance. The authors also thank to Mr. Fernando Mosos, Technical Director of Nuclear Affairs Division for his support.

Author details

Guillermo Parrado*, David Alonso, Julián Orozco, Mary Peña, Andrés Porras,
Martha Guzmán, Nelson Acero and Mauricio López

*Address all correspondence to: gparrado@sgc.gov.co

Colombian Geological Survey (Servicio Geológico Colombiano), Bogotá D.C., Colombia

References

[1] Instituto de Asuntos Nucleares. Ciencia y Tecnología para el progreso. 1st ed. Bogotá: Imprenta Nacional de Colombia; 1989. 269 p

[2] Castiblanco L. Parámetros Físicos del Reactor Nuclear IAN-R1. Nucleares. 1989;**4**(7):29-38

[3] Instituto de Ciencias Nucleares y Energías Alternativas. Libro de Registro de Operaciones del Reactor IAN-R1. Bogotá: INEA; 1997-1998

[4] Carvajal G, Dimitriadov A. Determinación de Yodo inorgánico estable en orina mediante Análisis por Activación. IAN. Bogotá; 1967. p. 21

[5] Arroyo A, Toro J, editors. Determinación de trazas de Selenio y Teluro en azufre por Activación Neutrónica. Bogotá: IAN; 1968. 13 p

[6] Arroyo A, Brune D, editors. Análisis por Activación de Uranio en rocas utilizando neutrones epitérmicos. Bogotá: IAN; 1972. 7 p

[7] Kunzendorf H, Løvborg L, Christiansen E, editors. Automated Uranium Analysis by Delayed-Neutron Counting. R-429 ed. Roskilde: RISØ; 1980. 38 p

[8] Instituto de Asuntos Nucleares. Informe de Actividades Técnicas y Administrativas durante el año 1973. Bogotá: IAN; 1974. 27 p

[9] Gomez H, Duque J. Análisis por Activación Neutrónica en la prospección de minerales. Nucleares. 1988;**3**(5):19-23

[10] Gomez H, Espinosa O. Análisis de Torio por Activación Neutrónica. Nucleares. 1986; **1**(1):3-8

[11] Instituto de Asuntos Nucleares. Informe de Labores 1989. Bogotá: IAN; 1989. 49 p

[12] Instituto de Asuntos Nucleares. Renovación del Reactor Nuclear IAN-R1. Santa Fé de Bogotá: IAN; 1992

[13] Greenberg R, Bode P, Fernandes E. Neutron activation analysis: A primary method of measurement. Spectrochimica Acta Part B. 2011;**66**:193-241

[14] Orvini E, Speziali M. Applicability and limits of instrumental neutron activation analysis: State of the art in the year 2000. Microchemical Journal. 1998;**59**:160-172

[15] Porras A. Importancia de la adecuación de un laboratorio de neutrones retardados en Colombia para la exploración de uranio y protección radiológica en su manipulación y análisis [thesis]. Universidad Nacional de Colombia – Sede Bogotá: 2016. 121 p. Available from: http://www.bdigital.unal.edu.co/53875/

[16] International Atomic Energy Agency. Normas de Seguridad del OIEA. Disposición final de desechos radiactivos Requisitos de Seguridad Específicos. SSR-5 ed. Vienna: IAEA; 2012

[17] Alonso D et al. Revisión sobre la Producción de Radionúclidos en Reactores Nucleares y sus aplicaciones como radiotrazadores. Revista Investigaciones y Aplicaciones Nucleares. 2017;**1**(1):6-23

[18] Bode P. Automation and quality assurance in the NAA facilities in Delft. Journal of Radioanalytical and Nuclear Chemistry. 2000;**245**:127-132

[19] Sierra O et al. Characterization of HPGe gamma spectrometric detectors systems for Instrumental Neutron Activation Analysis (INAA) at the Colombian Geological Survey. In: AIP Conference Proceedings. Vol. 1753. AIP; 2016. pp. 1-4

[20] Hamidatou L et al. Concepts, Instrumentation and Techniques of Neutron Activation Analysis [Internet]. 2013. Available from: https://www.intechopen.com/books/imaging-and-radioanalytical-techniques-in-interdisciplinary-research-fundamentals-and-cutting-edge-applications/concepts-instrumentation-and-techniques-of-neutron-activation-analysis [Accessed: May 2017]

[21] Norma ISO/IEC. Requisitos generales para la competencia de los laboratorios de ensayo y calibración. Bogotá: ICONTEC; 2005. 24 p

[22] Currie LA. Limits for qualitative detection and quantitative determination: Application to radiochemistry. Analytical Chemistry. 1968;**40**:586-593

[23] EURACHEM/CITAC. Quantifying Uncertainty in Analytical Measurement. Second Tech Rep, Guide CG4 s.1 ed. 2000

[24] International Organization of Standardization (ISO). Guide to the Expression of Uncertainty. Ginebra: ISO; 1993

[25] Oddone M, Meloni S, Genova N. Neutron activation analysis: A powerful tool for rare-earth elements assay in terrestrial materials. Inorganica Chimica Acta. 1984;**94**:141-146

[26] Orvini E, Speziali M, Salvini A, Herborg C. Rare earth elements determination in environmental. Microchemical Journal. 2000;**67**:97-104

[27] Norma ISO/IEC 17043. Evaluación de la conformidad — Requisitos generales para los ensayos de aptitud. Bogotá: ICONTEC; 2011. 28 p

[28] Coulomb R et al. Critical Minerals Today and in 2030: An Analysis of OECD Countries. Policy Paper, ESRC Centre for Climate Change Economics and Policy Grantham. United Kingdom: Research Institute on Climate Change and the Environment. Paris; 2015

[29] Rollinson H. Using Geochemical Data: Evaluation, Presentation, Interpretation. Prentice Hall; 1993. 352 p

[30] Organisation for Economic Co-operation and Development – OECD. Material resources, productivity and the environment: Key findings. 2013. 3-6 p

Chapter 4

Monte Carlo Simulation of Correction Factors for Neutron Activation Foils

Pham Ngoc Son and Bach Nhu Nguyen

Additional information is available at the end of the chapter

http://dx.doi.org/10.5772/intechopen.76984

Abstract

In the conventional neutron activation analysis, the elemental concentrations are normally determined from the comparison ratios between the measured specific activities of the sample and the standard reference material. An advantage in the comparison ratio method is that the systematic error due to neutron self-shielding and multi-scattering effects is canceled out, and the correction factors can be ignored but the preparation of reference standards to match the same conditions with those of various samples is the main difficulty. In the modern trend of neutron activation analysis, the K_0-standardization method has been developed and applied in almost all the NAA laboratories. An important research work in the procedure under this method is the characteristic information regarding the neutron source, such as thermal and epithermal neutron fluxes, and epithermal spectrum shape-factor. These neutron spectrum parameters are experimentally determined by using the activation foils, in which the corrections for all neutron effects cause systematic errors should be taken into account. Using the MCNP5 code, a well-known Monte Carlo simulation program, the results of correction factors of thermal, epithermal and resonance neutron self-shielding factors for Au, Co, Mn, W activation foils are presented in this chapter.

Keywords: neutron activation analysis, self-shielding factor, Monte Carlo simulation

1. Introduction

Neutron activation analysis is often performed with a reactor neutron spectrum. When the size of the irradiation sample or activation foils are not thin enough for ignoring the variance of neutron flux distribution, the information of the thermal, epithermal and resonance self-shielding effects should be considered for correction. The neutron capture reaction rate R(E), at neutron energy E per one atom of the irradiating sample in a neutron flux distribution $\phi(E)$, is defined as the following expression:

$$R(E) = \phi(E)\sigma_{\gamma}(E), \quad (1)$$

where σ_{γ} (E) is capture cross section. During the irradiation processes, strong resonance reactions deplete the neutron spectrum at the resonance energy due to absorption and scattering. For the energy regions just lower than the resonance energies, the neutron distribution is raised due to multiple scatterings from the resonances. Therefore, the capture reaction rate is affected because of the perturbation in the neutron spectrum. It is assumed that, for a sample or monitor with similar thickness but infinitely diluted, the neutron distribution inside this sample or monitor is a none-perturbed spectrum. The self-shielding correction factors can be estimated as the ratio of the reaction rate by the none-perturbed to that by the perturbed spectra.

The problem of calculations for determining the neutron self-shielding correction factors has been considered in neutron capture experiments, but in neutron activation analysis it is still less of information and numerical data. These correction factors are always needed to be taken into account in the data analysis of experiments such as neutron reaction cross sections measurements, neutron flux and spectrum measurements, and neutron activation analysis, and so on. In these experimental studies, the thickness of the irradiated sample or standard monitor is frequently not thin enough for ignoring the variance of neutron flux distribution inside the sample space. Therefore, the correction factors for the self-shielding and multi-scattering effects of thermal, epithermal and resonance neutrons should be determined exactly. This research topic had been previously carried out, and reported in case by case, which can be briefly discussed as follows. Lopes et al. [1] calculated the values of epithermal neutron self-shielding factors, including isotopic scattering, for foil of Au-197 and Co-59. Eastwood et al. [2] reported experimental values of resonance neutron self-shielding factors for foils and wires of Co-59, by the activation technique. Brose [3] measured the resonance neutron absorption factors for Gold foils with different thickness. Hisashi Yamamoto and Kazuko Yamamoto [4] reported their calculated values of resonance neutron self-shielding correction factors for foils of Au-197, W-186, Mn-55 and In-115. The effects of Doppler broadening and potential scattering were taken into account, considering only main individual resonances. Senoo et al. [5] introduced a Monte Carlo code, TIME-MULTI, for neutron multiple scattering calculations with time-of-flight spectra. Shcherbakov and Harada [6] proposed a fast analysis method for calculations of epithermal neutron self-shielding factors, which was made used the Padé approximation for Doppler broadening function. Trkov et al. [7] introduced a computer program for self-shielding factors in neutron activation analysis, in which the calculation method is based on the neutron slowing-down equation. The program can be used for calculations with multi-element samples and applicable only for reactor isotropic neutron field. Gongalves et al. [8, 9] performed resonance neutron self-shielding factors for foils and wires of different materials by using the MCNP code and proposed universal curves for a number of neutron source geometries, and applicable only for single element samples.

Although the research topic on neutron self-shielding has been considered for a long history, the available resources for neutron activation analysis (INAA) applications are still limited in both respects of numerical data and computational tools. The previous published values, calculations and measurements are mainly for the cases of resonance neutron self-shielding, and information is still lack for the cases of thermal and epithermal neutron self-shielding that are need in INAA with different standard foil and sample materials. In this chapter, the Monte Carlo code MCNP-5 has been applied for calculation of the thermal, epithermal and resonance neutron self-shielding factors for several standard neutron activation monitors that are often used in neutron activation measurements.

2. K_0-standardization method in neutron activation analysis

In the absolute standardization method, the concentration of the nuclide in a given sample can be determined as the following formula [10]:

$$\rho(\mu g/g) = N_p \left[\Delta_k . \frac{N_A . W . \theta . \gamma_k}{M} . \varepsilon_p . \Phi_e . \sigma_0 . \left(G_{th} . f + G_{epi} . Q_0(\alpha) \right) \right] . 10^6 \tag{2}$$

In the K_0-standardization method, the concentration of the nuclide in a given sample can be determined as the following formula [10]:

$$\rho_a(\mu g/g) = \left[\frac{\left(\frac{N_p/t_m}{w.S.D.C} \right)_a}{A_{sp,Au}} \right] . \frac{1}{k_{0,Au}(a)} . \frac{\left[G_{th} . f + G_{epi} . Q_0(\alpha) \right]_{Au}}{\left[G_{th} . f + G_{epi} . {}^{epi}Q_0(\alpha) \right]_a} . \frac{\varepsilon_{p,Au}}{\varepsilon_{p,a}} . 10^6 \tag{3}$$

in which G_{th} and G_{epi} are the thermal and epithermal self-shielding correction factors and the k0-factor is defined as [10]:

$$k_{0,c}(s) = \frac{M_c . \theta_s . \sigma_{0,s} . \gamma_s}{M_s . \theta_c . \sigma_{0,c} . \gamma_c} \tag{4}$$

3. Determination of α in the $E^{-(1+\alpha)}$ epithermal neutron spectrum

In a nuclear research reactor, the distribution of epithermal neutrons per unit energy interval is considered inversely proportional to the neutron energy. However, this assumption is only valid if the following conditions are satisfied [11].

- The medium in which fast neutrons are being slow down is homogeneous and infinite.
- The fast neutron sources are homogeneously distributed.
- The slowing down power is energy independent.
- There is no absorption during the moderation processes.
- The moderating atoms have the same mass as the neutron and behave as free particles.

In practical situations, these conditions are almost not satisfied in a nuclear reactor. Accordingly, the deviations from the 1/E distribution of epithermal neutron can occur in irradiation channels, and errors could be induced in the resonance integral defined as

$$I = \int_{E_{Cd}}^{\infty} \frac{\sigma(E)}{E} dE \tag{5}$$

where: E_{Cd} = 0.55 eV is the Cadmium cut off energy.

In order to take into account, the correction for deviation effect in the expression of the resonance integral, the $1/E^{1+\alpha}$ distribution has been introduced [11] where α is an energy-independent coefficient but its values dependent on the neutron source configuration. Therefore, the effective resonance integral is rewritten as follows:

$$I(\alpha) = \int_{E_{Cd}}^{\infty} \frac{\sigma(E)}{E^{1+\alpha}} dE \tag{6}$$

Experimental determination of the coefficient α for a specific neutron irradiation channel or facility is required for exact estimation of the corresponding effective resonance integral. The relationship between resonance integral I and effective resonance integral $I(\alpha)$ is expressed as the following equation [11]

$$I(\alpha) = (I - 0.426\sigma_0)\left(\overline{E_r}\right)^{-\alpha} + \frac{0.426\sigma_0}{2\alpha + 1} E_{Cd}^{-\alpha}, \tag{7}$$

where $\overline{E_r}$ is the effective resonance energy, and σ_0 is the 2200 m/s neutron capture cross section. The method for experimental instantaneous determination of α value, based on co-irradiation of three suitable resonance monitors, has been introduced by F. De CORTE [11], which represent as follows.

The specific count rate for an interesting γ-peak, emission from an irradiated sample, is defined as

$$A_{sp} = \frac{1}{m} \frac{C\lambda}{(1 - e^{-\lambda t_1})(e^{-\lambda t_2})(1 - e^{-\lambda t_3})}, \tag{8}$$

where t_1, t_2, t_3 are the irradiation time, decay time, measurement time, respectively, C the number of counts under γ-peak, and m the weight of irradiated sample. The specific count rate, A_{sp}, can be also calculated from the following expression:

$$A_{sp} = [f + Q(\alpha)]\phi_{epi}\sigma_0\varepsilon\gamma\theta C/M \tag{9}$$

where M, θ, γ, ε are atomic weight, isotope abundance, γ-ray absolute intensity, and the efficiency of the detector used in the gamma-ray spectrum measurement, respectively. $Q(\alpha)$ is the ratio of the resonance integral in the $1/E^{(1+\alpha)}$ epithermal neutron spectrum to the (n,γ) reaction cross section σ_0; f is the ratio of thermal to epithermal neutron flux [11].

$$Q(\alpha) = \frac{I(\alpha)}{\sigma_0} = (Q - 0.426)\overline{E_r}^{-\alpha} + \frac{0.426}{2\alpha + 1}E_{Cd}^{-\alpha} \tag{10}$$

$$\mathrm{Q} = \mathrm{I}/\sigma_0 \tag{11}$$

By using the co-irradiation of two suitable standard foils, denoted as 1 and 2, the flux ratio f can be determined by the following equation [11].

$$f = \left(\frac{\phi_{th}}{\phi_{epi}}\right)_{1,2} = \left[\frac{k_1\varepsilon_1}{k_2\varepsilon_2}Q_1(\alpha) - \frac{A_{sp,1}}{A_{sp,2}}Q_2(\alpha)\right]\left[\frac{A_{sp,1}}{A_{sp,2}} - \frac{k_1\varepsilon_1}{k_2\varepsilon_2}\right]^{-1}, \tag{12}$$

$$\mathrm{k} = \gamma\sigma_0\theta/\mathrm{M}. \tag{13}$$

When three resonance detectors (foils), denoted as 1, 2 and 3, are irradiated under the same experimental conditions, Eq. (12) can be rewritten for detector couples 1–2 and 1–3. Making equality between the quantities $f_{1,\,2}$ and $f_{1,\,3}$ leads to the following equation:

$$F(\alpha) = (a - b)Q_1(\alpha) - (a + 1)Q_2(\alpha) + (b + 1)Q_3(\alpha) = 0 \tag{14}$$

where

$$a = \left[\left(A_{sp,1}k_2\varepsilon_2\right)/\left(A_{sp,2}k_1\varepsilon_1\right) - 1\right]^{-1} \tag{15}$$

$$b = \left[\left(A_{sp,1}k_3\varepsilon_3\right)/\left(A_{sp,3}k_1\varepsilon_1\right) - 1\right]^{-1} \tag{16}$$

The coefficient α would be experimentally determined by solving the Eqs. (12) and (14).

In an ideal 1/E epithermal neutron spectrum, the resonance integral cross section for a neutron capture reaction is defined as follows:

$$I_0 = \int_{E_{Cd}}^{\infty} \frac{\sigma(E)}{E} dE, \tag{17}$$

where σ (E) is the neutron capture cross section as a function of neutron energy E, and E_{Cd} is the cadmium cut-off energy to be 0.55 eV when Cd shield thickness is 1 mm. In a non-ideal epithermal neutron spectrum which can be approximated by $1/E^{1+\alpha}$ distribution, the resonance integral is defined as:

$$I_0(\alpha) = \int_{E_{Cd}}^{\infty} \frac{\sigma(E)(1eV)^{\alpha}}{E^{1+\alpha}} dE, \tag{18}$$

The relation between I_0 and $I_0(\alpha)$ is defined as the following expression:

$$I_0(\alpha) = (1eV)^{\alpha}\left[\frac{I_0 - 0.426\sigma_0}{E_r^{\alpha}}\right] + \frac{0.426\sigma_0}{(2\alpha + 1)(E_{Cd})^{\alpha}}. \tag{19}$$

The experimental values of resonance integral $I_0(\alpha)$ for target x can be determined relative to that of $^{197}Au(n,\gamma)^{198}Au$ as a standard reaction by the following relation [11]:

$$I_0(\alpha)_x = I_0(\alpha)_{Au} \frac{(g\sigma_0)_x}{(g\sigma_0)_{Au}} \frac{(R_{Cd} - F_{Cd})_{Au}}{(R_{Cd} - F_{Cd})_x} \frac{G_{epi,Au}}{G_{epi,x}} \frac{G_{th,x}}{G_{th,Au}}, \tag{20}$$

The deviation of the epithermal neutron spectrum from the 1/E shape parameter can be experimentally determined by the 'Cd-ratio for multi-monitor' method, using the monitors of ^{197}Au, ^{59}Co, ^{186}W and ^{55}Mn. When a set of n monitors are irradiated with and without Cd-cover, the α-parameter can be obtained as the slope ($-\alpha$) of the straight-line $\log T_i$ versus $\log(E_{r,i})$ [11].

$$T_i = \frac{E_{r,i}^{-\alpha}}{(F_{Cd,i}R_{Cd,i} - 1)Q_{0,i}(\alpha)G_{e,i}G_{th,i}}, \tag{21}$$

where i denotes the ith isotope, F_{Cd} the cadmium transmission factor, R_{Cd} the Cd-ratio, Er the effective resonance energy in eV, G_{th} and G_{epi} the self-shielding factor for thermal and epithermal neutrons, and $Q_0(\alpha)$ is the ratio of the resonance integral in $1/E^{1+\alpha}$ epithermal neutron spectrum to the capture cross section σ_0 for 2200 m/s neutrons. The parameter T_i is a function of the α parameter, which can be determined by an iterative least square fit to the regression line.

4. Monte Carlo simulation method for neutron self-shielding calculations

Monte Carlo (MC) simulation is known as an essential numerical method for performing the statistical process of radiation interaction with material. The principle MC simulation in this subject is random selection of particle properties and its interaction behaviours from their probability distribution functions. By tracking the history of each particle during the interaction process, the information of particle fluxes, energy spectra and energy deposition in a specific cell of the simulating model can be obtained. Accordingly, the radiation dose rate at any position in the environment of the experiment can be estimated with statistical uncertainty. A typical block diagram for process of neutron particle transport are shown in the **Figure 1** [5].

5. Neutron self-shielding correction factors

In the neutron activation analysis experiments, the thickness of the samples and monitors may not be thin enough for ignoring the variance of neutron flux distribution and should be considered for correction. Therefore, the correction factors for thermal (G_{th}), epithermal (G_{epi}) and resonance (G_{res}) neutron self-shielding effects should be determined exactly. In this work, the Monte-Carlo code MCNP5 was used for calculations with specified case of irradiation foils with different thickness from 1×10^{-5} to 2 mm.

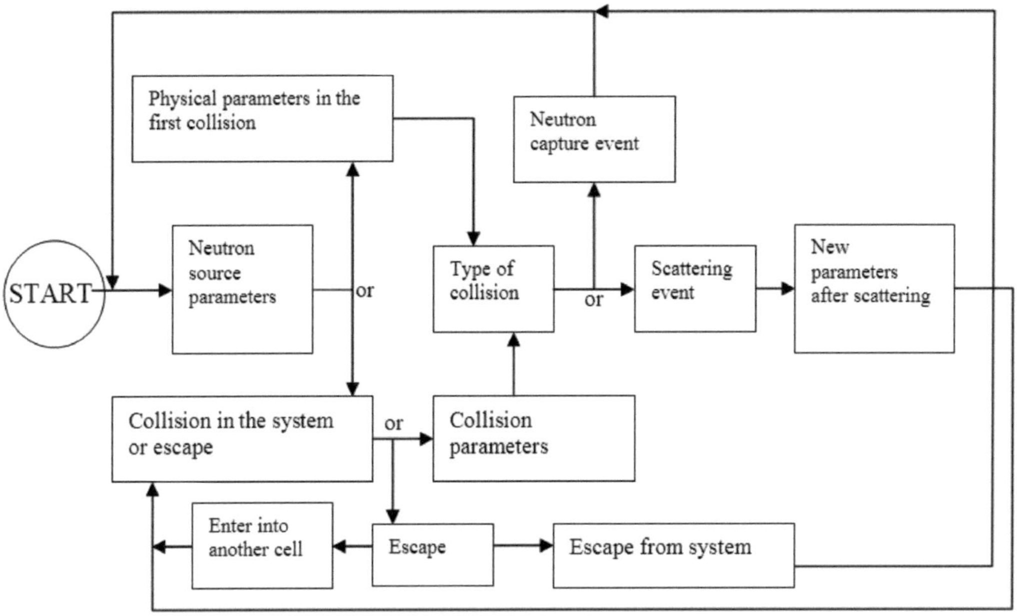

Figure 1. A typical block diagram for Monte-Carlo simulation of neutron transport process.

The neutron capture reaction rate R(E), at neutron energy E per one atom of the irradiating sample in a neutron flux distribution ϕ(E), is defined as the following expression:

$$R(E) = \phi(E)\sigma_\gamma(E), \tag{22}$$

where σ_γ (E) is capture cross section. During the irradiation processes, strong resonance reactions deplete the neutron spectrum at the resonance energy due to absorption and scattering. For the energy regions just lower than the resonance energies, the neutron distribution is raised due to multiple scatterings from the resonances. Therefore, the capture reaction rate is affected because of the perturbation in the neutron spectrum. It is assumed that, for a sample or monitor with finite thickness but infinitely diluted, the neutron distribution inside this sample or monitor is a non-perturbed spectrum. The correction factor a real sample with thickness 't' can be calculated as the following ratio.

$$G(t) = \int_{E_1}^{E_2} \varphi(E)\sigma_\gamma(E)dE / \int_{E_1}^{E_2} \varphi_0(E)\sigma_\gamma(E)dE, \tag{23}$$

where $\phi_0(E)$ is the original or non-perturbed neutron spectrum; ϕ(E) represents the perturbed neutron spectrum inside the real irradiating sample; E1 and E2 are, respectively, the lower and the upper limits of the neutron spectrum; for G_{th}: E1 = $1e^{-5}$ eV and E2 = 0.5 eV; for G_{epi}: E1 = 0.5 eV and E2 = 2e-1 MeV. The non-perturbed and perturbed neutron spectrum inside a real sample can be calculated by Monte-Carlo simulation using the MCNP5 code based on the ENDF/B-VII nuclear data library.

The results of MCNP5 simulations for perturbed and none-perturbed neutron spectra, $\phi(E)$ and $\phi_0(E)$, inside the real irradiating samples of Au-197 and Co-60 in different thickness is represented in **Figures 2** and **3**.

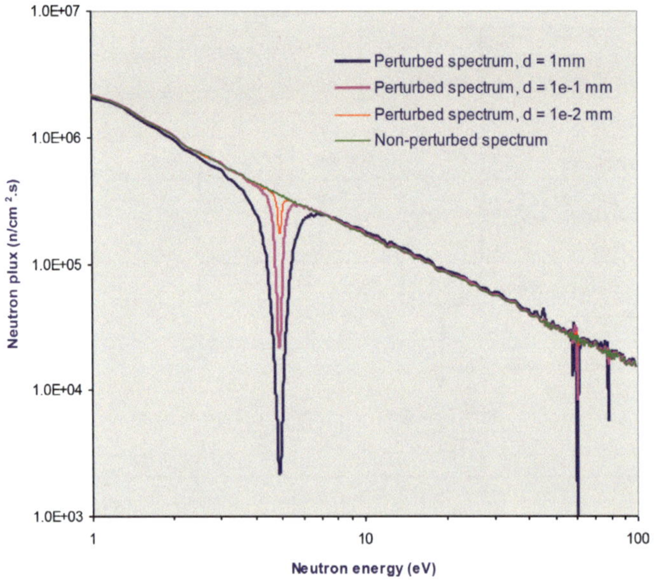

Figure 2. MCNP5 simulated results for perturbed and none-perturbed epithermal neutron spectrum in Gold foils of different thickness.

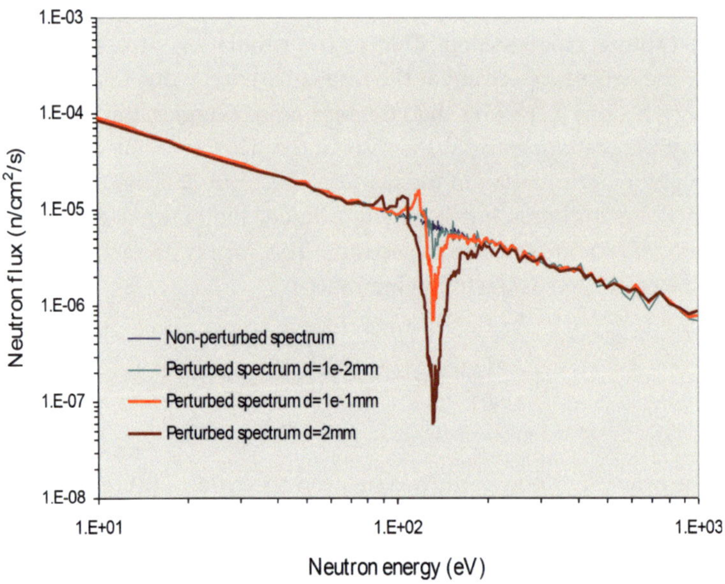

Figure 3. MCNP5 simulated results for perturbed and none-perturbed epithermal neutron spectrum in Cobalt foils of different thickness.

In order to make validation for the simulation procedure, the comparison between the results Gres factors with the literature data for foil materials of Au-197 and Co-59 have been carried out with different sample thicknesses as presented in **Figures 2** and **3**. Up to present, from the literature overview, the available data of neutron self-shielding factors are almost for the study of resonance neutron self-shielding correction G_{res} which is considered only in the energy region of primary resonance peak of reaction cross-section, but in NAA the thermal and epithermal neutron (or effective) self-shielding correction factor should be taken into account. As shown in **Figures 4** and **5**, the result of present simulated G_{res} factors for Gold and Cobalt foils has reasonable agreement with the experimental and calculated data by Gongalves [9], Brose [12] and Eastwood [2].

In order to provide effective information of neutron self-shielding correction factors in NAA, the validated simulation procedure has been applied for obtaining G_{th}, G_{epi} and G_{epi} factors for the activation nuclides of Au-97, Co-59, Mn-55 and W-186. In this work, the foil samples with 1.3 cm in diameter and thickness varying from 10^{-5} mm to 2 mm were used for simulations. The experimental configurations were simulated for the real condition of a reactor-based neutron activation experiment in which the irradiation channel can be described as a cylindrical isotropic neutron sources with sample flat to the channel axis. The dimensions of the irradiating channel are 30 cm length and 2.4 cm in diameter. These simulations were conducted for three case studies of neutron energy spectrum: (i) the Maxwellian distribution with average energy of 0.025 eV was

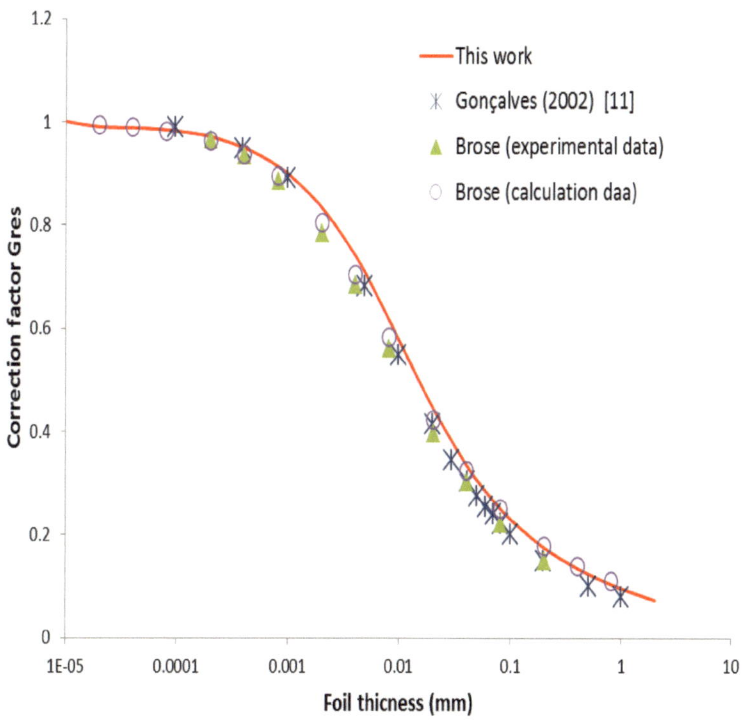

Figure 4. The calculated resonance neutron self-shielding correction factors for Gold foils of different thickness, in comparison with published data by Gongalves [9] and Brose [3].

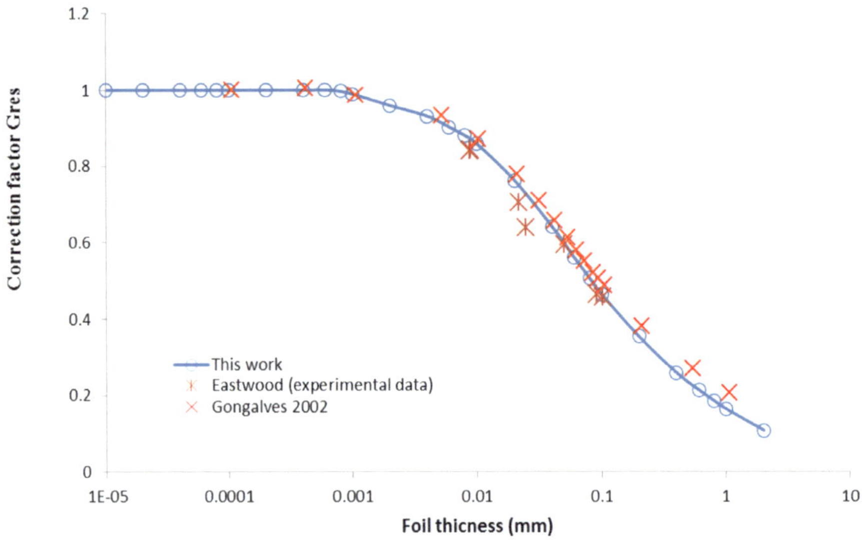

Figure 5. The calculated resonance neutron self-shielding correction factors for Cobalt foils of different thickness, in comparison with published data by Gongalves [9] and Eastwood [2].

applied for thermal neutrons, (ii) the pure 1/E distribution in energy range from 0.55 eV to 0.2 MeV was applied for epithermal neutrons, and (iii) the primary resonance energy peaks for resonance neutron spectrum. The results of thermal and epithermal neutron self-shielding correction factors for the activation foils of Au-197, W-186, Co-60 and Mn-55 in different thickness are presented in **Tables 1** and **2**.

Thickness (mm)	Neutron self-shielding correction factors for Au-197 foils			Thickness (mm)	Neutron self-shielding correction factors for Co-60 foils		
	Gth	Gepi	Gres		Gth	Gepi	Gres
0.00001	1	1	1	0.00001	1.000	1.001	1.000
0.00002	1	0.991	0.99	0.00002	1.000		1.000
0.00004	0.999	0.989	0.988	0.00004	1.000		1.000
0.00006	0.999	0.987	0.986	0.00006	1.000		1.000
0.00008	0.999	0.985	0.984	0.00008	0.999		1.000
0.0001	0.999	0.983	0.982	0.0001	0.999	1.001	1.000
0.0002	0.999	0.973	0.971	0.0002	0.999		1.000
0.0004	0.998	0.953	0.949	0.0004	0.998		1.000
0.0006	0.997	0.935	0.93	0.0006	0.997		1.000
0.0008	0.996	0.92	0.913	0.0008	0.997		0.997
0.001	0.996	0.906	0.898	0.001	0.996	0.998	0.989
0.002	0.993	0.846	0.834	0.002	0.995	0.996	0.958
0.004	0.988	0.76	0.74	0.004	0.992		0.930

Thickness (mm)	Neutron self-shielding correction factors for Au-197 foils			Thickness (mm)	Neutron self-shielding correction factors for Co-60 foils		
	Gth	Gepi	Gres		Gth	Gepi	Gres
0.006	0.984	0.696	0.671	0.006	0.989		0.902
0.008	0.981	0.646	0.618	0.008	0.987		0.880
0.01	0.977	0.607	0.575	0.01	0.985	0.992	0.858
0.02	0.961	0.486	0.445	0.02	0.974	0.988	0.762
0.04	0.934	0.384	0.334	0.04	0.957	0.981	0.642
0.06	0.912	0.335	0.282	0.06	0.941	0.977	0.562
0.08	0.893	0.306	0.251	0.08	0.928	0.974	0.506
0.1	0.876	0.286	0.229	0.1	0.916	0.971	0.465
0.2	0.805	0.234	0.175	0.2	0.865	0.961	0.355
0.4	0.707	0.194	0.135	0.4	0.792	0.946	0.259
0.6	0.636	0.175	0.116	0.6	0.736	0.935	0.214
0.8	0.58	0.162	0.104	0.8	0.690	0.925	0.186
1	0.535	0.152	0.096	1	0.651	0.915	0.164
2	0.388	0.124	0.073	2	0.516	0.873	0.108

Table 1. The results of neutron self-shielding correction factors foils of Au-197 and Co-60.

Thickness (mm)	Neutron self-shielding correction factors for Mn-55 foils			Thickness (mm)	Neutron self-shielding correction factors for W-186 foils		
	Gth	Gepi	Gres		Gth	Gepi	Gres
0.00001	1.000	1.000	1.000	0.00001	1.00	0.995	1.000
0.00002	1.000	1.000	1.000	0.00002	1.00	0.996	1.000
0.00004	1.000	1.000	1.000	0.00004	1.00	0.995	1.000
0.00006	1.000	1.000	1.000	0.00006	1.00	0.996	1.000
0.00008	1.000	1.000	1.000	0.00008	1.00	0.999	1.000
0.0001	1.000	1.000	1.000	0.0001	1.00	0.999	0.990
0.0002	1.000	1.000	1.000	0.0002	1.00	1.006	1.000
0.0004	1.000	1.000	1.000	0.0004	1.00	1.001	0.992
0.0006	1.000	1.000	1.000	0.0006	1.00	0.994	0.982
0.0008	1.000	1.000	1.000	0.0008	1.00	0.990	0.976
0.001	0.999	1.000	1.000	0.001	1.00	0.993	0.989
0.002	0.998	1.000	1.000	0.002	1.00	0.973	0.951
0.004	0.997	1.000	1.000	0.004	1.00	0.947	0.911
0.006	0.996	1.000	1.000	0.006	1.00	0.916	0.865
0.008	0.995	1.000	1.000	0.008	0.99	0.891	0.827

Thickness (mm)	Neutron self-shielding correction factors for Mn-55 foils			Thickness (mm)	Neutron self-shielding correction factors for W-186 foils		
	Gth	Gepi	Gres		Gth	Gepi	Gres
0.01	0.994	1.000	1.000	0.01	0.99	0.867	0.791
0.02	0.990	0.999	0.980	0.02	0.99	0.777	0.657
0.04	0.982	0.998	0.946	0.04	0.98	0.683	0.515
0.06	0.976	0.997	0.905	0.06	0.98	0.631	0.437
0.08	0.970	0.997	0.878	0.08	0.97	0.598	0.388
0.1	0.965	0.996	0.845	0.1	0.96	0.574	0.353
0.2	0.941	0.994	0.740	0.2	0.94	0.510	0.258
0.4	0.905	0.990	0.602	0.4	0.90	0.458	0.184
0.6	0.876	0.987	0.534	0.6	0.87	0.433	0.148
0.8	0.851	0.983	0.493	0.8	0.85	0.417	0.127
1	0.829	0.980	0.461	1	0.83	0.406	0.112
2	0.745	0.966	0.369	2	0.74	0.374	0.076

Table 2. The results of neutron self-shielding correction factors foils of Mn-55 and W-186.

6. Discussion and conclusion

Up to present, from the literature overview, the available data of neutron self-shielding factors are almost for the study of resonance neutron self-shielding correction G_{res} which is considered only in the energy region of primary resonance peak of reaction cross-section, but in NAA the thermal and epithermal neutron (or effective) self-shielding correction factor should be taken into account. Although the research topic on neutron self-shielding has been considered for a long history, the available resources for neutron activation analysis (INAA) applications are still limited in both respects of numerical data and computational tools. The previously published values, calculations and measurements, are mainly for the cases of resonance neutron self-shielding, and information is still lacking for the cases of thermal and epithermal neutron self-shielding that are needed in INAA with different standard foil and sample materials.

The MCNP5 code has been applied for exact simulation of neutron self-shielding correction factors of G_{th}, G_{res} and G_{epi} for activation foils of Au-97, Co-59, Mn-55 and W-186. The experimental configurations were simulated for the real condition of a reactor-based neutron activation experiment in which the irradiation channel can be described as a cylindrical isotropic neutron sources with sample flat to the channel axis. The three case studies of neutron energy spectrum: (1) the Maxwellian distribution with average energy of 0.025 eV was applied for thermal neutrons, (2) the pure 1/E distribution in energy range from 0.55 eV to 0.2 MeV was implemented for epithermal neutrons, and (3) the primary resonance energy peaks for resonance neutron spectrum. The results of simulated data with different foil thickness are presented in **Tables 1** and **2**.

http://dx.doi.org/10.5772/intechopen.76984

Author details

Pham Ngoc Son[1]* and Bach Nhu Nguyen[2]

*Address all correspondence to: pnson.nri@gmail.com

1 Nuclear Research Institute, Dalat, Vietnam

2 Representative Office of MIC in Da Nang City, Danang, Vietnam

References

[1] Lopes MDC, Avila JM. Energy dependence of neutron self-shielding factors in an isolated resonance. Nuclear Science and Engineering. 1990;**104**:40

[2] Eastwood TA, Werner RD. Resonance and thermal neutron self-shielding in cobalt foils and wires. Nuclear Science and Engineering. 1962;**13**:385

[3] Brose M. Zur Messung und Berechnung der Resonanzabsorption von Neutronen in Goldfolien. Nukleonika. 1964;**6**:134

[4] Yamamoto H, Yamamoto K. Self-shielding factors for resonance foils. Journal of Nuclear Science and Technology. 1965;**2**:421

[5] Senoo K, Nagai Y, Shima T, Ohsaki T. A Monte Carlo code for multiple neutron scattering events in a thick sample for (n, g) experiments. Nuclear Instruments and Methods in Physics Research A. 1994;**339**:556

[6] Shcherbakov O, Harada H. Resonance self-shielding corrections for activation cross section measurements. Journal of Nuclear Science and Technology. 2002;**39**(5):548-553

[7] Trkov A, Žerovnik G, Snoj L, Ravnik M. On the self-shielding factors in neutron activation analysis. Nuclear Instruments and Methods in Physics Research A. 2009;**610**:553-565

[8] Gongalves IF, Martinho E, Salgado J. Monte Carlo calculation of epithermal neutron resonance self-shielding factors in wires of different materials. Applied Radiation and Isotopes. 2001;**55**:447-451

[9] Gongalves IF, Martinho E, Salgado J. Monte Carlo calculation of epithermal neutron resonance self-shielding factors in foils of different materials. Applied Radiation and Isotopes. 2002;**56**:945-951

[10] Pomme S, Hardeman F, Robouch P, Etxebarria N. Neutron Activation Analysis with K0-Standardization: General Formalism and Procedure. SCK-CEN report. 1997; BLG-700

[11] Benjelloun M, Paulus JM. α-Determination in the $E^{-(1+\alpha)}$ epithermal neutron spectrum of the CNR Reactor (Strasbourg). Journal of Radioanalytical and Nuclear Chemistry. 1987; **109**(2):393-401

[12] X-5 Monte Carlo Team. MCNP - Version 5, Vol. I: Overview and Theory. LA-UR-03-1987. 2003

Section 2

Neutron Activation Systems and Technologies

Chapter 5

Neutron Activation System for ITER Tokamak

Vitaly Krasilnikov, MunSeong Cheon and
Luciano Bertalot

Additional information is available at the end of the chapter

http://dx.doi.org/10.5772/intechopen.75966

Abstract

Neutron activation system (NAS) is currently designed for ITER and will be procured by the Korean DA. The main purpose of this diagnostic is to evaluate the integrated fusion power and cross-check with other neutron diagnostic, whose sensitivity can vary over time. Total neutron production rate shall be measured from all over the plasma, regardless of the position or profile of the neutron source. Therefore, it is required to minimize material and its density variation across the field of view between the plasma and the irradiation end.

Keywords: neutron activation system, ITER, nuclear, fusion, tokamak, plasma, diagnostic

1. Introduction

Neutron activation system is a diagnostic measuring the absolute neutron flux and fluence on the first wall [1]. It utilizes pneumatic post method to send a sample of material close to the plasma, where it gets activated by neutrons. This sample is then retrieved back with the same pneumatic post technique and the activation of the sample is measured with gamma-gay spectrometers [2]. The main goal of the ITER neutron activation system (NAS) is to evaluate the total neutron production rate from all over the plasma. The measurement accuracy depends on the position and profile of the plasma and the material in front of the irradiation end. It is required to minimize the amount of material and its density variation across the field of view between the plasma and the irradiation end. Due to the radiation and thermal environment of the ITER in-vessel, however, the measurement from ITER NAS cannot avoid the strong influence from in-vessel materials such as the diagnostic first wall, blanket modules, and divertor cassettes that are located near the irradiation ends. A number of irradiation positions located above and below the plasma as well as on high-field side and low-field side has been selected for the ITER NAS to compensate the strong influence from in-vessel materials such as the diagnostic first wall, blanket modules, and divertor cassettes.

2. Generic description

ITER NAS measures gamma radiation from samples activated by fusion neutron flux. Encapsulated samples are transferred between irradiation ends and counting station by the driving of nitrogen (or helium) gas. Tubes of diameter 12.7 mm will be used for the transfer lines of the capsule.

NAS consists of the pneumatic transfer system and the counting system (**Figure 1**) Cheon et al. [3]. The pneumatic transfer system includes gas supply, transfer station, transfer line, irradiation ends, counter ends, and disposal bin. It is the subsystem related with the transfer of the encapsulated samples from the loading to the disposal. The PLC-based control system will be harnessed for the accurate operation of the system. The counting system consists of gamma-ray detectors, electronic devices such as high voltage supplies and amplifiers, and tool for neutron source strength evaluation. It is the system for the evaluation of the parameters of the NAS by counting gamma-rays from the activated samples [4].

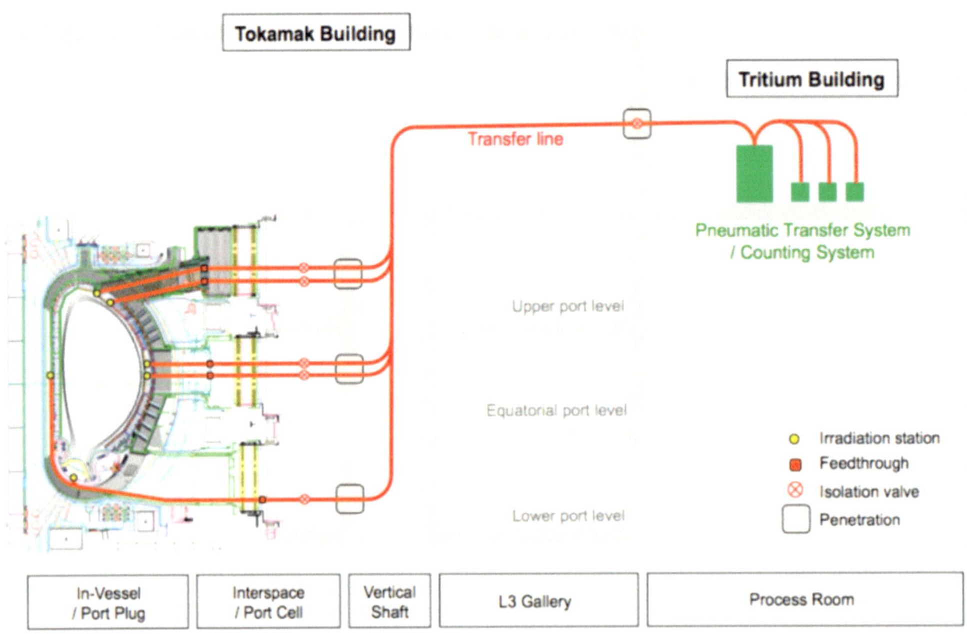

Figure 1. The scheme of neutron activation system for ITER.

Due to the large size and the elongated shape of ITER plasma, multiple positions for the irradiation ends in toroidal section are required for highly reliable measurements. At present, four irradiation end locations per toroidal section (A, B, C and D in **Figure 2**) are planned for ITER NAS considering reliability of the measurement and redundancy of the system.

Transfer tubes of the NAS should be bent many times to reach the irradiation ends from the transfer station. To avoid capsule stuck problem around tube bends, there should be a minimum bending radius of the tube in designing tube route. All bends of the tube should have larger radius than this minimum bending radius. Assuming the capsule of OD 8 mm and L 30 mm, and the tube of ID 9 mm, the minimum bending radius of the tube is about 100 mm. The current design value of the minimum bending radius is 150 mm, with the safety factor 50% applied.

Current port allocation for the NAS is #11 and #18 for the upper port, #11 and #17 for the equatorial port, and #12 and #18 for the lower ports. For points A and B, the irradiation ends will be located inside the port plugs. Other irradiation ends will be installed on the vacuum vessel wall with the pipelines routed through the lower level ports [5]. Allocated ports and port numbers for the irradiation locations are shown in **Figure 2**. Total number of the irradiation ends which will be installed is 12.

Transfer station distributes capsules to the appropriate locations such as irradiation end, counting station, or disposal bin. It consists of capsule loader and distribution machine 'carousel.' When capsule is loaded on the carousel from the loader, the platter inside the carousel rotates to place capsule to the point connected to the designated place. The capsule loader and the carousel are separated by the air lock system to prevent the leakage of the driving gas. At every transfer line ends the air cushion technique, which will be implemented to prevent capsule breakage.

Counting station locates outside the bioshield of ITER where neutron flux effect on the detectors is negligible. Detectors such as HPGe or NaI will be used to count gamma-rays from

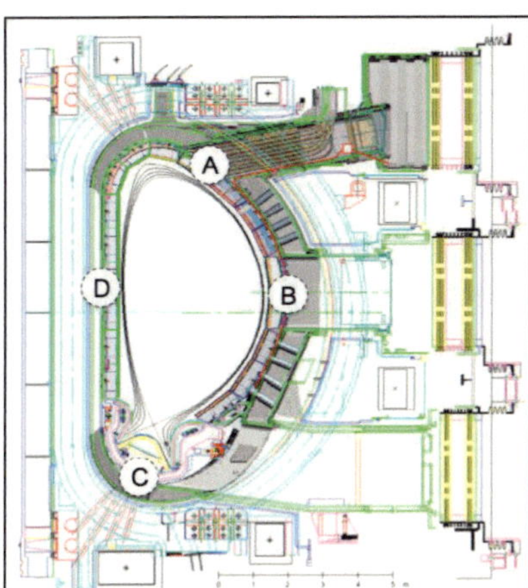

	Port	Port #
A	Upper	#11, #18
B	Equatorial	#11, #17
C	Lower	#12, #18
D	Lower	#12, #18

Figure 2. Distribution of irradiation ends in a toroidal section and allocated port numbers.

the activated samples. The required parameters for the NAS such as neutron fluence will be evaluated from the gamma spectrum considering the location of the irradiation end, sample material and its mass, and irradiation and cooling time.

3. Design of DFW cutout for upper port irradiation end

NAS is supposed to provide reliable and robust measurement data because it will be used for the calibration of other neutron diagnostics. From the point of reliability and robustness of the measurement, optimum location of the irradiation end is where the activation coefficient is insensitive to any environmental changes during the plasma operation and measurement, such as geometrical change of the surrounding material, plasma movement, and slight movement of irradiation end location. The geometrical changes of the irradiation end surrounding material can be happened due to the thermal expansion, vibration, distortion, and so on. Thus, location far away from plasma without any scattering material can be the best place for the irradiation end.

However, materials between the plasma and irradiation end cannot be avoided in real situation. If the location of the irradiation end is far away from the plasma, too much material in-between will increase the measurement uncertainty. On the other hand, if the location of the irradiation end is very close to the plasma, plasma movement will increase the measurement uncertainty as well. So we should find a location where the effect of the plasma movement and the effect of the material are the modest. Normally, an irradiation end without any surrounding material nearby is chosen as the location in given position (by the port location, for example). If the effect of the plasma movement is very significant, compensation of the measurement can be necessary: (1) by using plasma location information from other diagnostics or (2) by measuring simultaneously in the opposite location vertically or radially.

However in ITER, where the radiation environment is extremely harsh, it is very difficult to avoid material around the irradiation end. Instead, we will try to find geometry of the surrounding material, whose impact on the measurement is minimized, with the help of neutron transport calculation.

The irradiation end in the upper port is selected as the object of the investigation because it is one of the locations inside the port plug, where the effect of the geometry change of the surrounding material is less severe than other locations. Most of the in-vessel irradiation ends are located between the blanket shields, where is vulnerable to the geometrical change. The activation coefficients of various samples with and without DFW material have been compared around the irradiation end (see **Figure 4** for instance). The effect of the geometry of the cutout in DFW was investigated to find a design: (1) whose absolute value of the activation coefficient is similar with the one without DFW material and (2) whose response to the plasma movement is not so severe.

Activation coefficients of three samples, that is, silicon, copper, and titanium at the upper port irradiation end were calculated using FISPACT and MCNP code. Objective nuclear reactions are $^{28}Si(n,p)^{28}Al$, $^{63}Cu(n,2n)^{62}Cu$, and $^{48}Ti(n,p)^{48}Sc$.

http://dx.doi.org/10.5772/intechopen.75966

Figure 3 shows the MCNP model for the calculation. The cutout of DFW was designed to have a toroidal and poloidal angle of view as large as possible, while minimizing the amount of material in front of the irradiation end to the plasma direction, in order to minimize errors from the plasma movement and neutron transport calculation. Initial values for each dimension are:

- Depth: 130 mm.
- Poloidal angle: 105°.
- Toroidal angle: 60°.
- Toroidal width: 30 mm.

Calculated activation coefficients are shown in **Figure 4**. When there is no DFW material (upper line) and when there is a cutout in DFW material (lower line). Absolute values of the activation coefficient are reduced by about 10% when the irradiation end is surrounded by

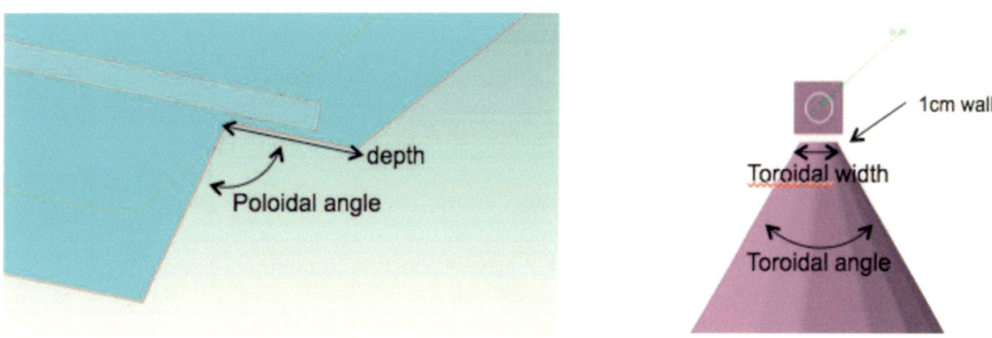

Figure 3. MCNP model for calculation: (left) side view and (right) front view.

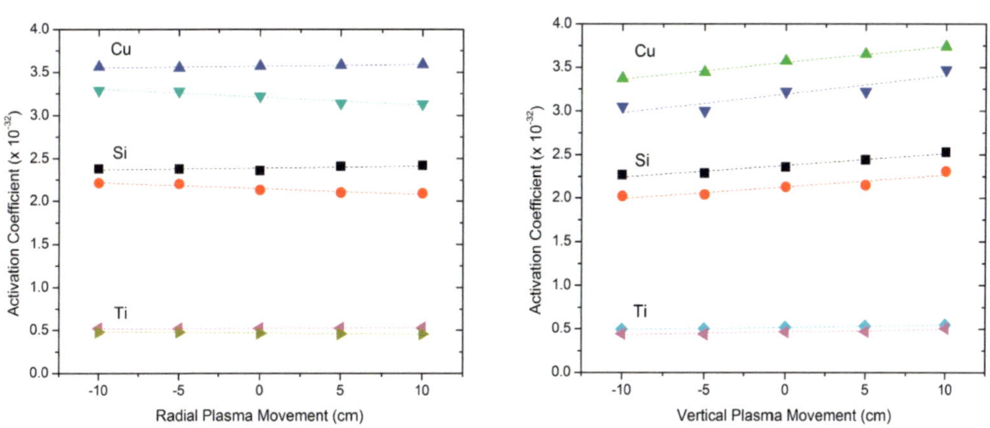

Figure 4. Comparison of plasma movement effect with and without DFW.

DFW material. In spite of the DFW surrounding the response of the irradiation end to the vertical movement of plasma is almost the same with the one without DFW except for the absolute value shift. However, clear decrease of the activation coefficient can be identified when plasma moves outward radially. This can introduce additional error about 2.5% by ±10 cm radial movement of plasma.

Effect of the toroidal width of the cutout was investigated, and the result is shown in **Figure 5**. The width was increased from the initial value (30 mm) up to the geometrical maximum (170 mm) and the activation coefficient of $^{63}Cu(n,2n)^{62}Cu$ reaction was investigated by moving the plasma position in the radial direction. The absolute values of the activation coefficients become closer as the width of the cutout increases. The differences between the 'No-DFW material' case are about 10, 2, and 0.8%, when the widths are 30, 100, and 170 mm, respectively, when the plasma is kept at its central place. Also the response to the plasma

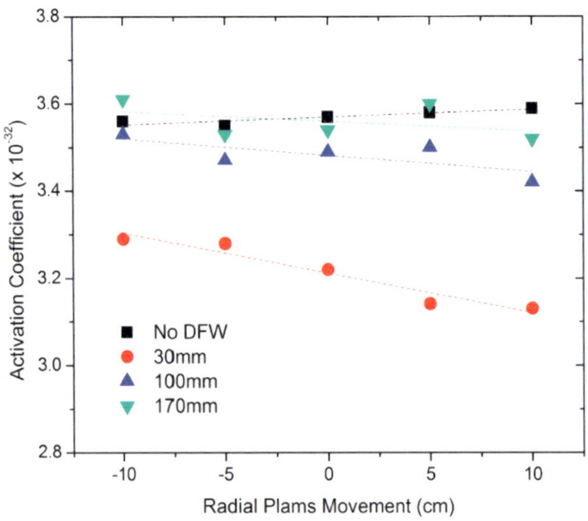

Figure 5. Toroidal width effect on activation coefficients of 63Cu(n,2n)62Cu response by radial plasma movement.

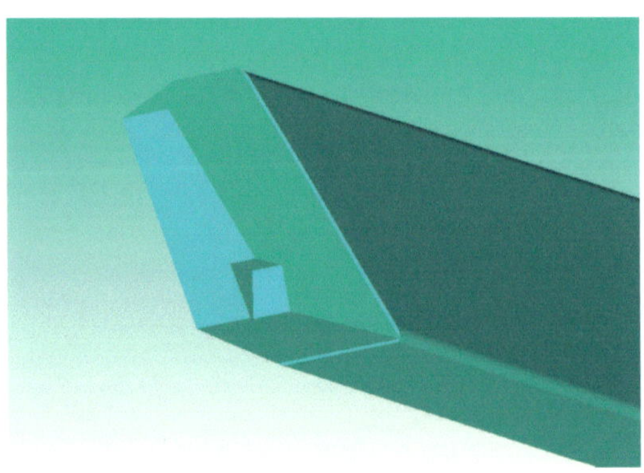

Figure 6. Image of DFW cutout for NAS.

movement is improved by increasing the width. It is easily identified the response of the irradiation end become more insensitive to the plasma movement as the size of the width increases. Slops of the linearly fit equations of the calculated activation coefficients are 9.2, 3.8, and 2.2 ($\times 10^{-34}$) per 1 cm plasma movement, when the widths are 30, 100, and 170 mm, respectively. Calculated maximum errors according to this equation are 0.8 and 1.4%, when the plasma movement values are ±5 cm and ± 10 cm, respectively.

The effect of the DFW cutout design on the measurement accuracy was investigated. The initial design values are proved to be proper except the toroidal width. It is recommended the toroidal width of the cutout to be as large as possible. The recommended design of the DFW cutout is shown in **Figure 6**. By making a cutout according to the design recommended by this calculation, we can imitate as much as possible the response of the ideal irradiation end, where there is no surrounding material nearby.

4. Evaluation of measurement accuracy

Measurement accuracy of NAS with 12 irradiation ends is estimated using MCNP calculations. The response of each irradiation location is evaluated by changing the location and the profile of the neutron source (see **Figure 7**).

The evaluated result of the neutron source displacement effect (**Figure 8**) shows that the upper port is the best position for the irradiation due to its lowest sensitivity. The induced error due to the vertical displacement can be even lower when it is compensated with the measurement at divertor position, as long as the irradiation end at divertor is well characterized during the plasma operation. It is estimated that induced error from the neutron source displacement can be ~ ± 1% even without compensation from other diagnostics, from the simultaneous

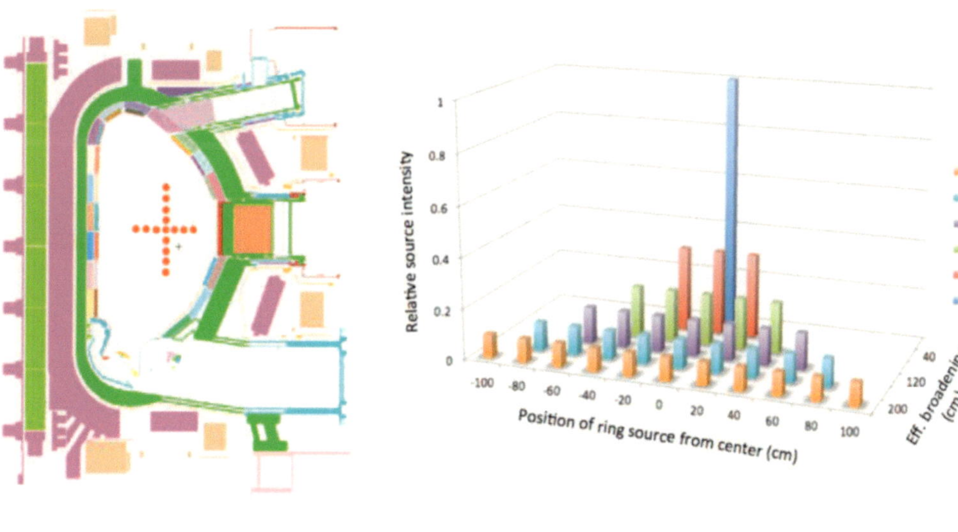

Figure 7. Evaluation of the effect of neutron source position and broadening.

measurement from the upper and divertor position, when the displacement range is within ±20 cm vertically and radially. The equatorial port position can be used for backup when the data are compensated from other diagnostics.

The effect of neutron source broadening (**Figure 9**) on the measurement, which cannot be estimated during the in-vessel calibration, was evaluated. The result also indicates that the upper port is the best position because it has the lowest effect from the neutron source broadening, and shows good characteristic of depending only on the vertical broadening. It is interesting to note that the equatorial port position shows symmetric measurement with the upper port position. Therefore, the simultaneous measurements from the upper and equatorial port position are expected to provide the total neutron production with the broadening error of ~1% without compensation from other diagnostics, when the profile peaking factor is in the range of $3 < \alpha < 7$.

The calculations show that with the combination of the measurements from the upper port, equatorial port, and divertor region can provide relatively good evaluation of the total neutron production in the plasma. In spite of the low reliability of the measurement from the inboard midplain position, it is reasonable to keep this irradiation ends, as they are the only ones capable of providing the absolute value of the neutron flux coming to the inboard side.

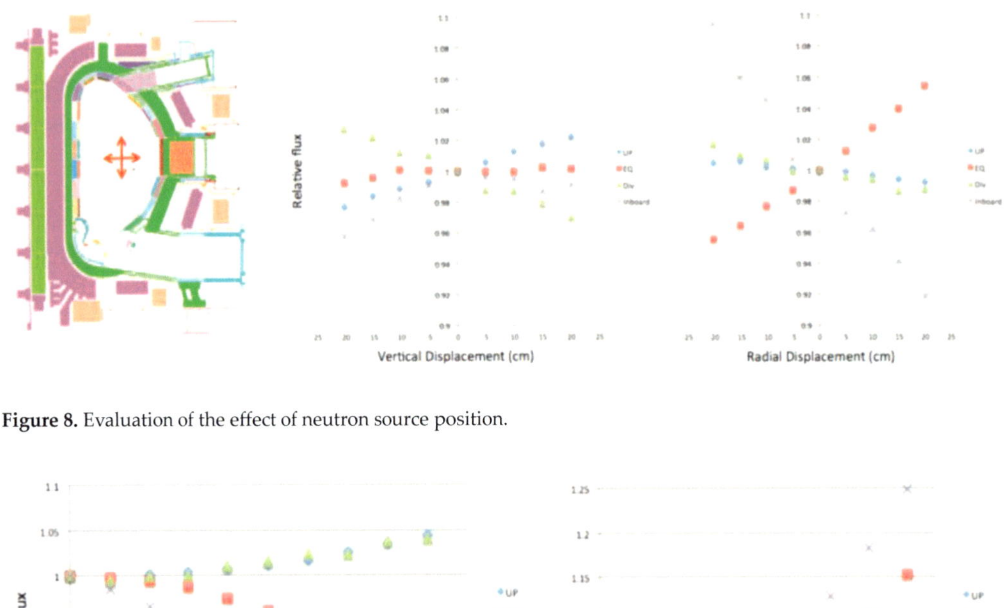

Figure 8. Evaluation of the effect of neutron source position.

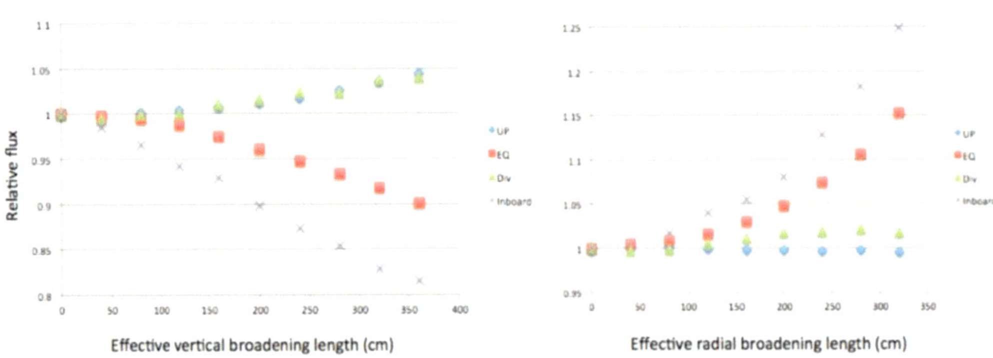

Figure 9. Evaluation of the effect of neutron source broadening.

5. Designs of the NAS components for ITER

Thermal analysis has made significant impact on the design of the NAS front-end components (**Figure 10**). All NAS components installed inside the vacuum vessel shall follow the design guideline SDC-IC (Structural Design Criteria for ITER In-vessel Components), which requires the maximum temperature of the components to be less than about 500°C. According to the simple thermal analysis on the irradiation end in the upper port, the temperature of the irradiation end is found to exceed 500°C when the irradiation end protrudes only by 6 cm from the actively cooled diagnostic shield module (DSM) inside (but not touching) the diagnostic first wall (DFW) that has a full depth of 60 cm. Similarly, all in-vessel irradiation ends located inboard side of the vacuum vessel are found to exceed 500°C, when there is no active cooling of the irradiation end structures. The temperature could be below 500°C only when the forced circulation of He gas with the velocity higher than 10 m/s is provided for the in-vessel transfer line during the plasma operation, which can be problematic when the gas blowing with such velocity fails, for example, when the capsule touches the irradiation location and plugs the hole for the gas circulation. In order to resolve the thermal issue, the design is updated to cool down all in-port irradiation ends by attaching the cooling jacket around the irradiation end structure, where coolant can be supplied from the in-port coolant manifold.

Port plug irradiation ends mainly consist of two transfer lines which are composed of coaxial or parallel tubes (**Figure 11**). Most components will be fabricated with SS316L except the capsule monitoring cabling, which consists of MgO mineral insulated (MI) cables and Al_2O_3-based electrical feedthrough. The front part of the irradiation end is enclosed with the coolant housing, which is connected with the coolant tubing. Two guiding rings are attached on the outside of the coolant housing for the smooth insertion of the irradiation end into the DSM. The feedthroughs will be welded on the closure plate of the port plugs.

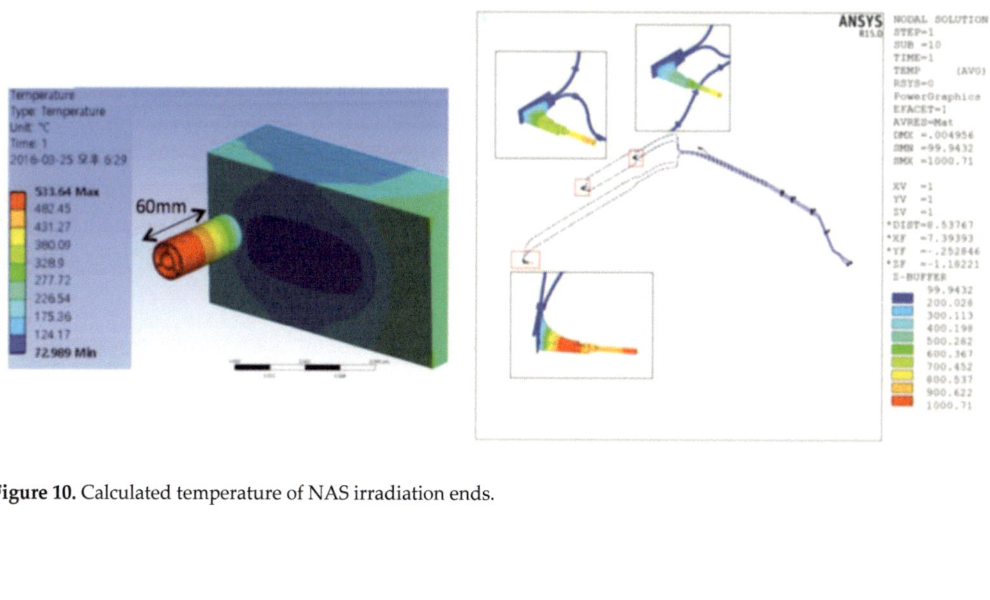

Figure 10. Calculated temperature of NAS irradiation ends.

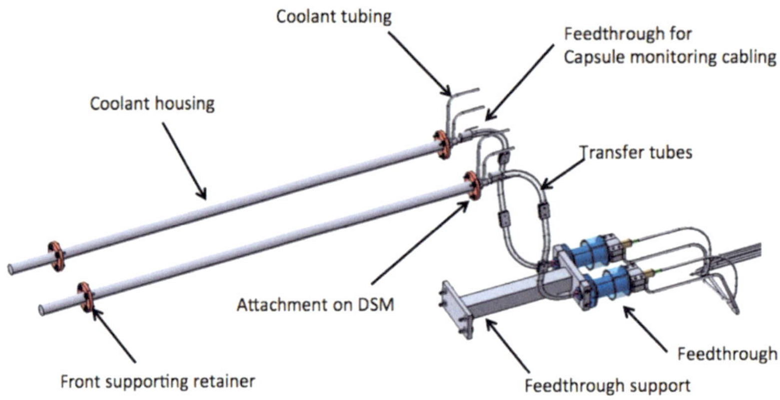

Figure 11. Port plug transfer line in EP11.

Transfer station consists of many moving components such as a servo-motor, linear actuators and many solenoid or gas-driven valves. Pneumatic properties of the transfer system for transferring capsule are as below:

- Pressure of gas in reservoir: ~8 bars max.
- Pressure of driving gas: 1–8 bars
- OD of sample transfer tube: 12.7 mm
- Thickness of sample transfer tube: 1.25 mm
- OD of retrieving gas tube: 12.7 mm
- Thickness of retrieving gas tube: 1.25 mm
- Diameter of capsule: ~8 mm
- Length of capsule: ~20 mm

Samples will be transferred to the designated position by the action of distribution machine 'carousel' (**Figure 12**). Rotating platter inside the carousel will transfer sample to the loading position which are connected to the designated position. When the samples are ready, the valves behind are opened to shoot them to the designated positions. Before arriving at the designated position, the speed of them will be slowed down to prevent breakage. A Programmable Logic Controller (PLC) will control the operation of the transfer system. Figure 9.4.2 is a current design of the carousel.

Counting station measures gamma-rays from the activated samples. It consists of gamma-ray detector, signal processing electronics such as high voltage supply, preamplifier, amplifier, and multichannel analyzer, and data analyzing software. Many gamma-ray detectors such as gas chambers, scintillators, and semiconductor detectors are commercially available. Among these detectors, NaI detectors and HPGe detectors are the most commonly used

http://dx.doi.org/10.5772/intechopen.75966

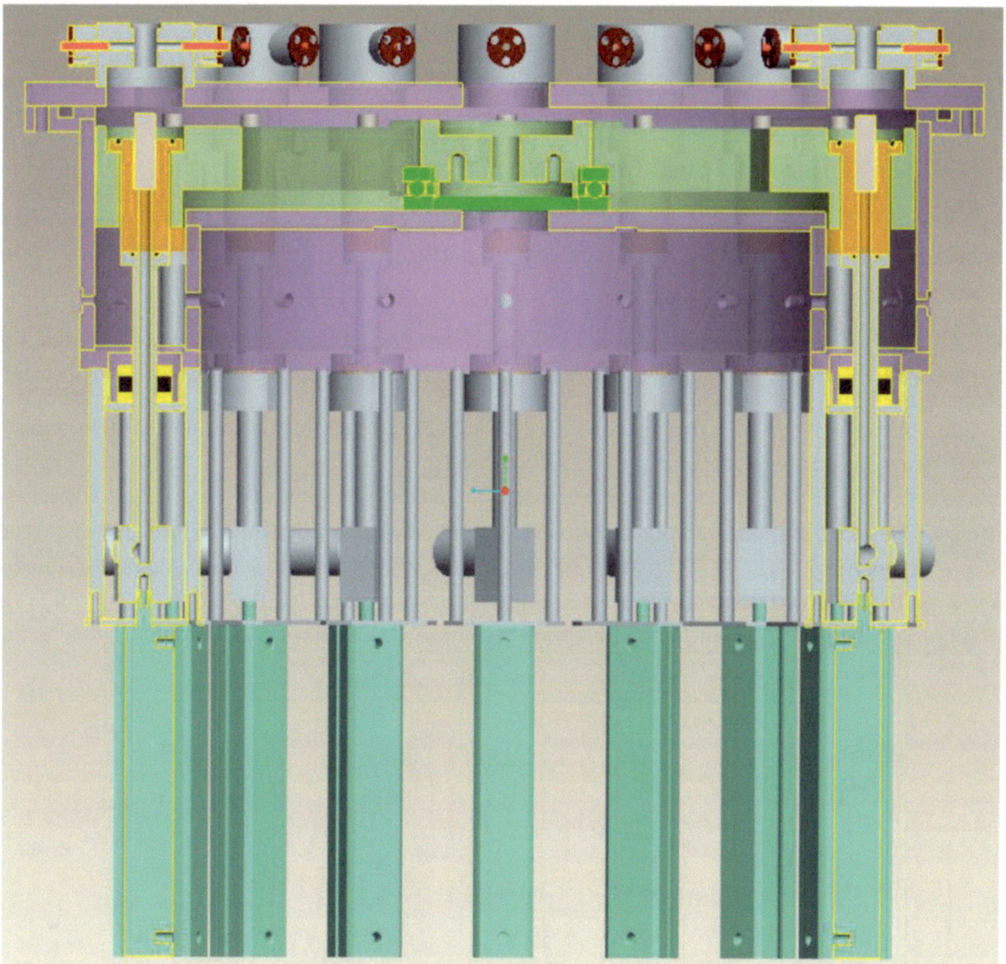

Figure 12. Design of sample distribution machine 'carousel'.

ones for neutron activation analysis, but other types of detector can be also considered. Appropriate detectors will be chosen for the proper operation of ITER NAS considering state of the art.

6. Performance assessment

The NAS system has been designed for determining the total neutron yield during the DT operation. The system must provide also time-resolved measurements of the global neutron source strength and evaluation of the fusion power. Measurement of absolutely calibrated neutron flux and fusion power will be performed.

Various tools are used for carrying out the analysis:

- MCNP v.5 (Monte Carlo N-Particle) transport code is used for the calculation of neutron fluxes and neutron energy spectra at the designated locations for the irradiation.
- FENDL-2.1 (Fusion Evaluated Nuclear Data Library) is used as the material database for the MCNP calculation
- FISPACT-2007 is used for the inventory of neutron induced activation of the sample materials.
- EAF-2007 (European Activation File) is used for the source of cross-section data for FISPACT-2007
- Lite series (A-lite, B-lite, and C-lite) 40° sector ITER geometrical model with a fusion plasma neutron source is used for the MCNP calculation.

An irradiation location at midplane inboard region is selected for the calculation of neutron flux and spectrum with MCNP code. The flux of this location is the second strongest among seven poloidal irradiation locations. Two tallies are designated for the irradiation ends, one is very close to the first wall, and the other is behind the blanket module very close to the

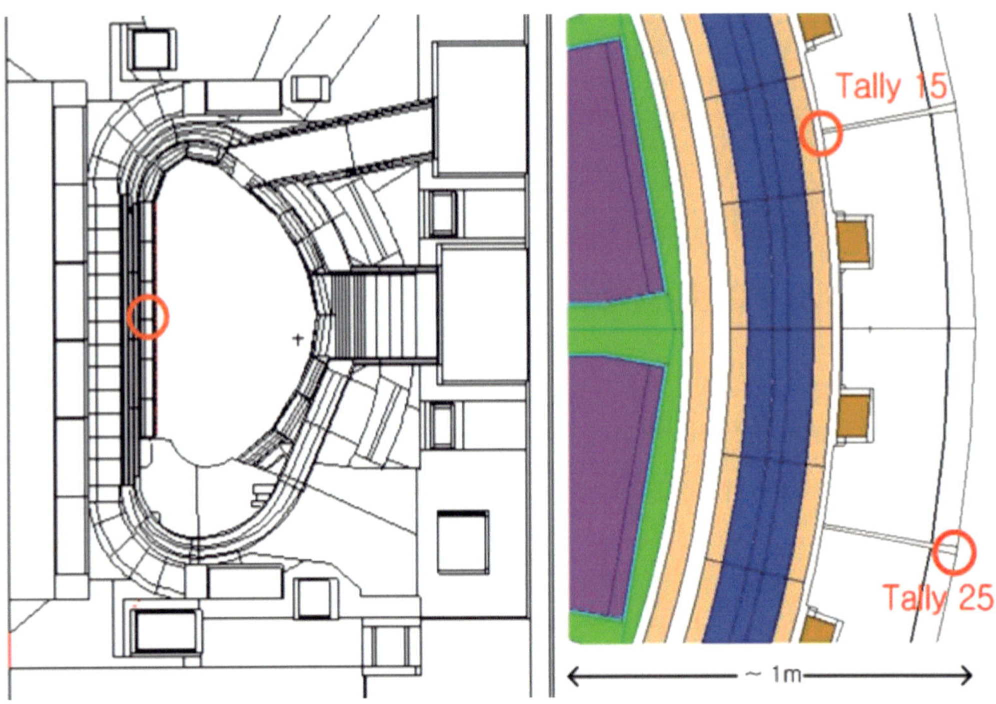

Figure 13. Tally locations in the Alite model.

http://dx.doi.org/10.5772/intechopen.75966

vacuum vessel wall. Both are located at the cross point of horizontal and vertical gap centers of four blanket modules. Tally 15 is located near the inner VV wall whereas Tally 25 is facing the plasma. **Figure 13** shows two tally locations. Si, Al, Ti, Fe, Nb, and Cu have been selected as sample materials for the investigation [6]. Samples are assumed to be a foil type with the diameter of 7 mm and the thickness of 0.1 mm.

Figure 14 shows the calculated neutron spectra at two tallies. Total neutron fluxes at tally 15 and tally 25 are 5.45 × 1013 and 5.9 × 1014 s^{-1} cm^{-2} respectively, assuming 500 MW of fusion power. In spite of the heavy blanket modules structure in front of the irradiation end, the spectrum of tally 15 shows clear 14 MeV neutron peak. This is due to the blanket modules acting as a collimator that absorbs scattered neutrons. Calculated neutron flux and spectrum are used for input data of FISPACT for the calculation of the sample activity.

As one of the requirements of the ITER NAS is to measure time-integrated neutron fluence to the first wall for all discharge duration, it is desirable for samples to be irradiated as long as possible time within the discharge time. Thus, the activities of various samples are calculated with the irradiation of 1000 s, and the result is shown in **Figure 15** D-T fusion power of 500 MW is assumed for the flux calculation.

Another requirement is to provide supplementary neutron flux data with a crude temporal resolution of about 10 s, when necessary for a backup or calibration of other flux measurement systems, such as Microfission Chambers (MFC) and neutron flux monitors (NFM). Thus, the activities of various samples are calculated with the irradiation of 10 s, and the result is shown in **Figure 16**.

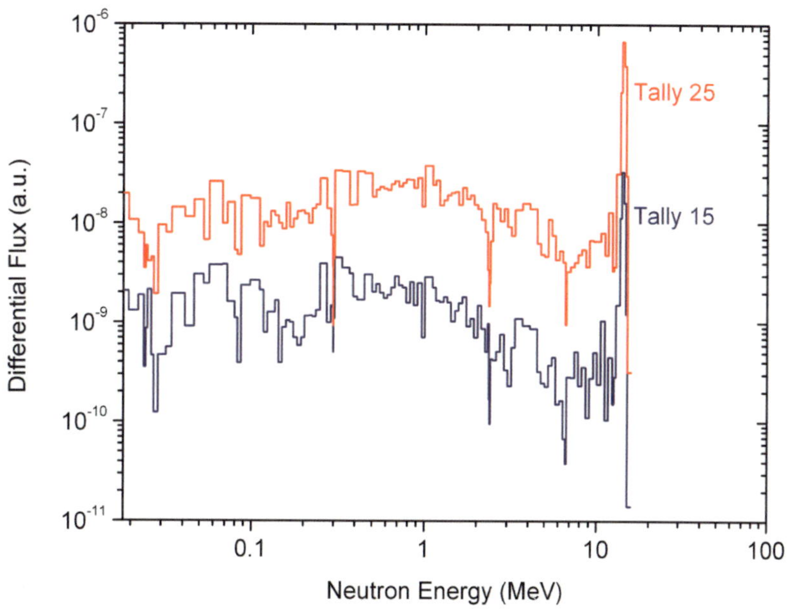

Figure 14. Neutron spectra at tally 15 and tally 25.

The activation desired for a sample should be similar to that provided by a standard source used for absolute calibration of the gamma-ray detectors. A typical maximum value for modestly safe handling would be 100 μCi. **Figure 17** shows the fusion power needed to create 100 μCi samples assuming 10-s irradiation and 20-s cooling at a irradiation location D.

Assuming the mass of samples to be from a few milligrams to a few grams, the fusion power that NAS can cover ranges from a few hundred watts to gigawatts by using various sample materials at different irradiation end locations. This measurement range satisfies the required measurement range both of the neutron flux and the fusion power.

Figure 18 shows the fusion power needed to create 100 μCi samples assuming 1000-s irradiation and 1000-s cooling at an irradiation location D. This result also shows that the NAS can measure neutron fluence in a long pulse operation condition of ITER. Si is not an appropriate sample material for the long time irradiation because the activity of Si saturates when the irradiation is much longer than the half-life of Si.

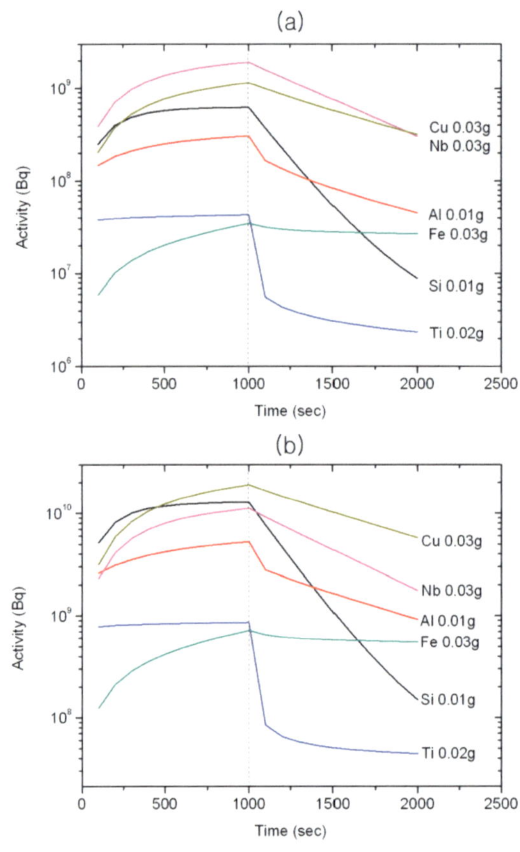

Figure 15. Activation by irradiation of 1000 s (a) tally 15, (b) tally 25.

http://dx.doi.org/10.5772/intechopen.75966

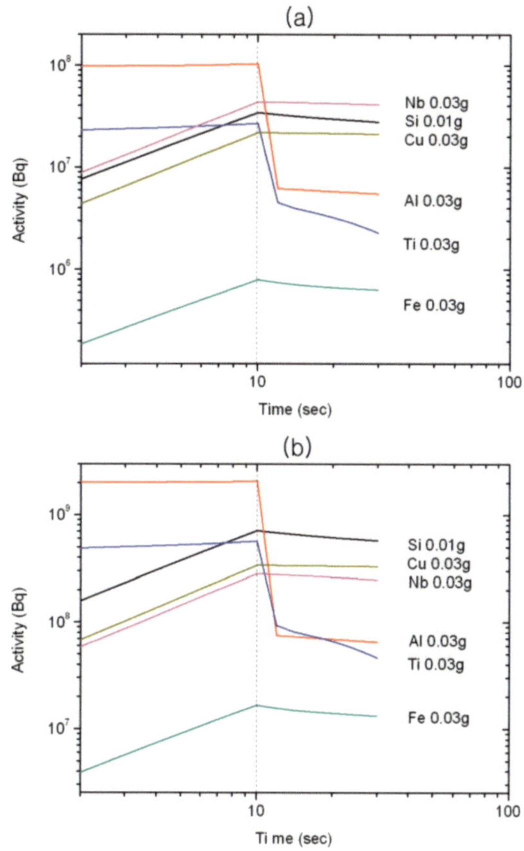

Figure 16. Activity by irradiation of 10 s (a) tally 15, (b) tally 25.

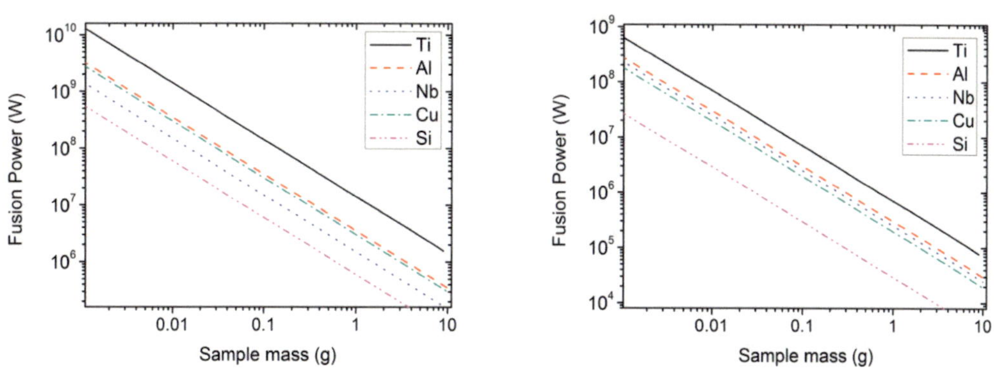

Figure 17. Fusion power needed to create 100 μCi samples by the 10-s irradiation and 20-s cooling (left) at tally 15, and (right) at tally 25.

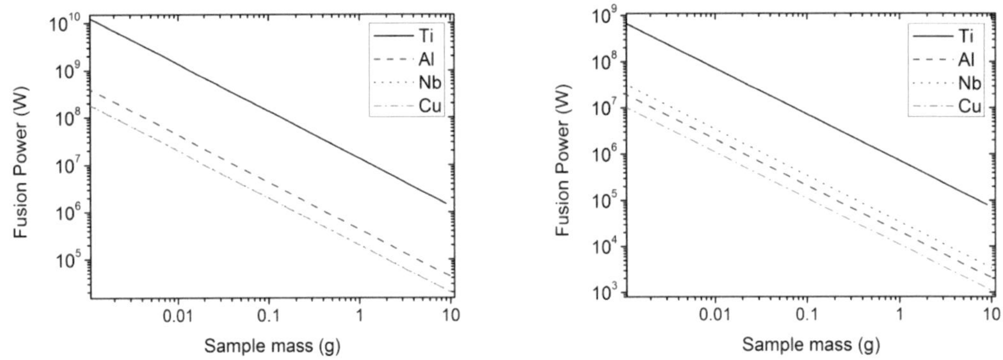

Figure 18. Fusion power needed to create 100 μCi samples by the 1000-s irradiation and 1000-s cooling (left) at tally 15, and (right) at tally 25.

7. Summary

The ITER neutron activation system that has been briefly presented in the earlier sections is under development by the Korean Domestic Agency of ITER. Despite the challenges driven by ITER's unprecedented thermal, electromagnetic and nuclear loads, those driven by high activation in full-power operation leading to very limited personnel access and the highest safety and reliability requirements [7], despite all these aspects, the presented NAS design proves to be suitable to satisfy ITER's measurement requirements.

Author details

Vitaly Krasilnikov[1,3]*, MunSeong Cheon[2] and Luciano Bertalot[1]

*Address all correspondence to: vitaly.krasilnikov@iter.org

1 ITER Organization, St. Paul-lez-Durance, France

2 National Fusion Research Institute, Daejeon, Republic of Korea

3 Tokamak Energy Ltd, Abingdon, UK

References

[1] Krasilnikov A et al. Status of ITER neutron diagnostic development. Nuclear Fusion. 2005;**45**:1503-1509

[2] Cheon MS, Pak S, Lee HG, Bertalot L, Walker C. In-vessel design of ITER diagnostic neutron activation system. Review of scientific instruments. 2008;**79**(10):10E505. DOI: 10.1063/1.2990857

[3] Cheon MS, Seon CR, Pak S, Lee HG, Bertalot L. Development of the prototype pneumatic transfer system for ITER neutron activation system. Review of Scientific Instruments. 2012;**83**(10):10D303. DOI: 10.1063/1.4729673

[4] Barnes CW, Loughlin MJ, Nishitani T. Neutron activation for ITER. The Review of Scientific Instruments. 1997;**68**:1-577. DOI: 10.1063/1.1147657

[5] Encheva A, Bertalot L, Macklin B, Vayakis G, Walker C. Integration of ITER in-vessel diagnostic components in the vacuum vessel. Fusion Engineering and Design. 2009;**84**:736-742

[6] Lee Y, Dang J, Jo J, Chun K, Hwang Y, Cheon M, Lee H, Bertalot L. Preliminary study on capsule material for ITER neutron activation system. Fusion Engineering and Design. 2014;**89**(9-10):1894-1898

[7] Vayakis G, Hodgson ER, Voitsenya V. Chapter 12: Generic diagnostic issues for a burning plasma experiment. Fusion Science and Technology. 2008;**53**:699-750

Chapter 6

An Overview of the Establishment of Methodology to Analyse up to 5g-Sample by k_0-Instrumental Neutron Activation Analysis, at CDTN, Brazil

Maria Ângela de B.C. Menezes and
Radojko Jaćimović

Additional information is available at the end of the chapter

http://dx.doi.org/10.5772/intechopen.83812

Abstract

The team of the Laboratory for Neutron Activation Analysis, Brazil, has been continuously improving the k_0-instrumental neutron activation analysis, the k_0-INAA method, having Jožef Stefan Institute, Slovenia, as partner researcher of the neutron activation technique. The group aims at answering the analytical requests of customers and the needs of the researches developed by the lab. The latest improvement was to establish a methodology to analyse up to 5 g-samples. The usual procedure in neutron activation analysis is to determine elemental concentrations in small samples of about 200 mg, a geometrical point source. The reason why these samples are used is that this geometry brings about a number of simplifications during irradiation and gamma spectrometry. This paper describes the steps carried out in the development of the large sample methodology that has already been published elsewhere and has been applied successfully. The results of some reference materials and samples are displayed. It is important to mention that this research has confirmed that any other laboratory applying k_0-INAA is able to establish this methodology without having to modify its facilities, since the neutron self-shielding, gamma attenuation, and detector efficiency over the volume source are established.

Keywords: k_0-instrumental neutron activation analysis, large sample, detector efficiency, neutron self-shielding, gamma-ray attenuation

1. Introduction

When an analytical result is available, several tasks had been accomplished before such as the routine procedure establishment, calibration of instruments, quality assurance and quality control (QA/QC), training of technical group, good laboratory practice and others. All of these requirements are needed to meet the demand for analytical values that are even increasing in all fields, contributing to environmental monitoring, individual and population health and economical decisions. According to the development of human knowledge, the requirements for quality have been diversified and increasing in several fields. Therefore, the demand for analytical data with smaller uncertainties and lower detection limits is expanding and more requirements are necessary to meet the quality required by the clients.

The nuclear analytical technique, NAA [1, 2], fulfils several requirements. It is well-known that NAA requires a non-chemical preparation—a non-destructive technique—and analyses a large number of elements simultaneously. Besides, it is a traceable technique [3, 4]. It presents sensitivity, multi-element ability, selectivity and versatility and determines chemical elements with precision and accuracy [5, 6]. That is why it is a powerful technique.

The technique is well established at the Laboratory for Neutron Activation Analysis, LNAA, located at the Nuclear Technology Development Centre (CDTN) sponsored by the Brazilian Commission for Nuclear Energy (CNEN), in Belo Horizonte, capital of the Brazilian state of Minas Gerais. The nuclear research reactor, the 100 kW TRIGA MARK I IPR-R1, has enabled the NAA to be applied determining the elemental concentration of different samples, such as soil, sediment, plants, food, medicines and biological tissues of humans and animals, among others [7–20]. The NAA has been applied through relative and parametric methods and has been applied meeting requests of customers both of CDTN and at industries, universities and other institutions. The technique has also been applied in researches of the LNAA. The standardised k_0-method [21] was established in 1995, being the most efficient form of application of this nuclear analytical technique [8, 9]. The k_0-method has been continuously improving along its nuclear data [22, 23], which can be found in the form of an Excel file, the k_0-database 2015 [24].

2. Small samples versus large samples

The usual procedure in NAA is to analyse a sample whose mass is lower than 500 mg, considering it as a geometrical point source. This entails a number of simplifications during irradiation and gamma spectrometry [25, 26]. This way, several simplifications can be made such as disregarding the neutron self-shielding, neutron-flux gradients over the sample and self-attenuation of gamma rays. The impact to the accuracy of the results is negligible.

On the other hand, there is a growing demand for the NAA established at CDTN to explore its potential in order to overcome the main limitations when analysing point samples, which are: to reach lower detection limits than those currently in use (for instance in food samples, plants, medicines and lichens) and to carry on analysis at lower cost, that is, to analyse a smaller number of samples and shorter time of analysis. For example, instead of analysing 20 small samples, a single about 4 g composite sample could be analysed; to provide greater representativeness of samples of non-homogeneous materials, for instance, industrial waste materials; to enable the analysis of whole parts in which it is not possible or permitted to remove an aliquot for analysis, for example, of archaeological ceramics. In addition, the low neutron flux of a low-power reactor can be compensated by increasing the amount of sample to be exposed to irradiation.

A possibility to overcome these problems is to analyse larger samples—samples of more than 0.5 g [25–32]. In order to obtain reliable analysis results, some parameters should be determined: (i) detector efficiency evaluation over the volume source, (ii) neutron flux depression due to absorption and scattering and (iii) the relative attenuation of gamma rays originating from different positions within the sample.

During the irradiation, the neutron field is perturbed during absorption and scattering inside the sample. It is called neutron self-shielding. This can be overcome by experimentally determining the neutron flux distribution in real samples in a defined volume for a matrix [6, 33, 34]. The degree of gamma self-attenuation depends on a number of factors such as sample geometry, linear attenuation coefficient, material density, sample composition and photon energy [35].

The laboratories that have been applying the neutron activation to large samples (LS-INAA) analyse samples in a range of kilogrammes, and for this procedure, special facilities are required, for the activation as well as for the detection. For instance, in Delft, The Netherlands, a facility was built to irradiate and measure samples from 2 to 50 kg [3, 26–31].

3. Development of the methodology applied at CDTN

The LNAA determines chemical elemental concentrations following the usual procedure—small cylindrical samples. The irradiations are carried out in the carousel facility of the TRIGA MARK I IPR-R1 reactor that operates at 100 kW with an average thermal neutron flux of 6.3×10^{11} cm^{-2} s^{-1}. The laboratory has a high demand of analysis, answering the clients' request, analysing several kinds of samples. It is often necessary to overcome the difficulties due to low neutron fluency, inhomogeneity of unknown sample and time consumption of analysis. For that reason, to analyse larger samples would be an attractive possibility. However, it is not allowed to change the infrastructure of irradiation; therefore, a study was developed to verify the possibility to analyse 5 g-samples, maximum mass content in the irradiation vial, 25 times larger than usual samples analysed. The k_0-method of neutron activation analysis [21] would be applied and the current infrastructure for irradiation and gamma spectrometry facilities would be used. To develop this study, the mass of the small sample analysed was around 200 mg and the larger cylindrical sample, around 5 g.

Aiming at solving the main limitations when dealing with small samples and exploring a new possibility of analysis, a methodology of analysing larger samples or cylindrical samples was established in LNAA at CDTN [16, 36, 37]. All experiments were developed in geological matrix. The reason was that matrix is the one most used in routine elemental analysis at CDTN. All irradiations were performed in the carousel facility of the 100 kW TRIGA MARK I IPR-R1 reactor under an average thermal neutron flux of 6.3×10^{11} cm^{-2} s^{-1} and average spectral parameters f (thermal to epithermal ratio neutron fluxes) and α (the epithermal flux distribution parameter), 22.67 and 0.0026, respectively. The induced activities were measured on absolutely calibrated HPGe detectors [9]. The vials were inserted in a polystyrene container for irradiation during 8 hours.

Gamma spectroscopy was carried out after suitable decay on an absolute calibrated HPGe detector named D4, GC 5019, CANBERRA, with 50% relative efficiency. The absolute calibration was done following a recommended procedure in the k_0-standardisation method [6, 9, 21].

The HyperLab programme [38] was used for peak area deconvolution and the software package Kayzero for Windows® [39], also called KayWin, was applied. It is a specific programme to calculate the elemental concentration for the so-called small samples—routine procedure—including the efficiency and coincidence correction calculations. It also calculates the values of efficiency for each energy based on the experimental full-energy peak efficiency determined previously for point-source geometry aiming at point sample analysis. This software determines the reference full-energy peak efficiency, ε_p, of the detector [9].

The main experimental steps carried out to establish the methodology based on Menezes and Jaćimović [16] and Menezes et al. [36, 37] will be described. Some tables and figures related to the development have already been published and are shown here with permission. However, other tables are original.

To establish the methodology to analyse a large sample, it was necessary to check three parameters: (i) detector efficiency over the volume source, (ii) neutron self-shielding during neutron irradiation and (iii) gamma-ray attenuation within the sample during counting [26, 35, 40].

3.1. Detector efficiency over the volume source

To evaluate the detector efficiency over the volume of the sample applying the k_0-method, using the KayWin software, it is necessary to determine the reference full-energy peak efficiency, ε_p. This programme calculates the elemental concentration for small samples—routine procedure—while calculating the efficiency and coincidence correction. The values of efficiency for each energy are also calculated based on the experimental full-energy peak efficiency. This efficiency was determined previously for point-source geometry aiming at point sample analysis. It is necessary to provide the detector characteristics, container and geometry dimensions, composition of the sample and reference efficiency curve to determine its efficiency. The final calculations will give the full-energy peak efficiency, ε_p, and the effective solid angle ($\overline{\Omega}_{eff}$).

The KayWin software calculates, for each gamma energy in a spectrum, the efficiency, when the elemental concentration of a real sample is determined. The detector efficiency can also be determined by ANGLE V3.0 software [41] that was successfully installed at CDTN. This software is specific to calculate the full-energy peak efficiency of the semiconductor detector to several source geometries as point and cylindrical shapes, Marinelli, etc. In this study, it was used to validate the full-energy peak efficiency, εp, calculated by KayWin. Both software need information on the detector characteristics, container and geometry dimensions, composition of the sample and reference efficiency curve. The full-energy peak efficiency, ε_p, and the effective solid angle will be calculated. It is important to mention that ANGLE software makes efficiency calculations only for coincidence-free gamma lines.

3.2. Neutron self-shielding and the spatial neutron flux distribution factors

In this step, the neutron self-shielding and the spatial neutron flux distribution factors [36, 40, 42–44] during irradiation in geological matrix were determined. The objective was to evaluate how significant were their contributions to the final elemental concentration results and which correction factor may be determined to correct the neutron flux gradient and self-shielding effects.

3.2.1. Neutron self-shielding

Several programmes calculate the thermal neutron self-shielding factor (G_{th}). To calculate this factor, it is necessary to know the sample composition and geometry. In this methodology development, the KayWin [39] and MATSSF software [45] were applied.

3.2.2. Spatial neutron flux distribution factors

It is relevant to verify the axial and radial distributions of neutron fluxes in the vial. They can be evaluated experimentally during irradiation [40] and after by measuring the $F_{c,Au}$-factor, called the comparator factor, calculated based on neutron monitors irradiated together with the sample in sandwich form. This factor provides the trend of the axial neutron flux gradient, while radial gradient is negligible due to similar diameter of the sample and standard (Al-0.1% Au alloy).

Simplified equation of the k_0-method [21] for elemental concentration calculation for an analyte (a), is below, Eq. (1):

$$\rho_a = \frac{A_{sp,a}}{F_{c,Au}} \cdot \frac{1}{k_{0,Au}(a)} \cdot \frac{1}{G_{th,a}.f + G_{e,a}.Q_{0,a}(\alpha)} \cdot \frac{1}{\varepsilon_{p,a}} \tag{1}$$

where the $F_{c,Au}$-factor, the comparator factor, Eq. (2), is:

$$F_{c,Au} = \frac{A_{sp,m}.10^{-6}}{k_{0,Au}(m)} \cdot \frac{1}{G_{th,m}.f + G_{e,m}.Q_{0,m}(\alpha)} \cdot \frac{1}{\varepsilon_{p,m}} \tag{2}$$

where $A_{sp,m}$ and $A_{sp,a}$ are the specific activities of monitor (m) and analyte (a); $k_{0,Au}(m)$ and $k_{0,Au}(a)$ are k_0-factors of monitor Au (by definition ≡ 1) and analyte; $\varepsilon_{p,m}$ and $\varepsilon_{p,a}$ are the full-energy peak detection efficiency of the monitor (Al-0.1%Au alloy in disc form) and radionuclide of analyte; f is thermal to epithermal neutron flux ratio; $G_{th,m}$ and $G_{th,a}$ are the correction factors for thermal neutron self-shielding; $G_{e,m}$ and $G_{e,a}$ are the correction factors for epithermal neutron self-shielding; $Q_{0,m}(\alpha)$ and $Q_{0,a}(\alpha)$ are resonance integral ($1/E^{1+\alpha}$) to 2200 m s^{-1} cross-section ratio; α is the epithermal flux distribution parameter.

Note that the $F_{c,Au}$-factor, as defined in Eq. (2), is proportional to the epithermal neutron flux density and directly indicates a gradient in epithermal flux density. In this study, the $F_{c,Au}$-factor was calculated by KayWin software based on several Al-0.1%Au monitors irradiated together with the samples in sandwich form.

3.3. Gamma-ray attenuation within the sample during counting

The k_0-method of NAA requires a precise technique to calculate full-energy peak detection efficiency (ε_p) of an HPGe detector for diversified samples and various counting geometries. The procedure provided by Moens et al. [46] presented a semi-empirical approach that has been introduced in KayWin software via option SOLCOI.

The basic principles are:

a. A "reference" counting geometry is chosen to measure a set of calibrated point sources at a large distance, e.g., 15–20 cm;

b. For a sample, the $\varepsilon_{p,geom}$ for particular gamma energy is expressed by employing the concept of the effective solid angle ($\overline{\Omega}_{eff}$), with "*ref*" denoting the reference geometry and "*geom*" the actual one, Eq. (3):

$$\varepsilon_{p,geom} = \varepsilon_{p,ref} \cdot \frac{\overline{\Omega}_{geom}}{\overline{\Omega}_{ref}} \tag{3}$$

It is important to mention that by the definition of the effective solid angle, two factors are included: (i) the attenuation effects which gamma rays undergo outside an HPGe detector active zone, F_{att}-factor and (ii) the probability for an energy degradable gamma-ray interaction with the detector material, F_{eff}-factor. Both factors can be calculated analytically.

4. Experimental steps and results

In this study, the reference material IAEA-SOIL-7 (International Atomic Energy Agency (2000) [47] was analysed as a small cylindrical sample (SS), ~200 mg, and as a large cylindrical sample (LS), ~5 g. Neutron flux monitors and Al-0.1% Au disc, IRMM-530R (6 mm diameter-disc and 0.1 mm high) were used.

To carry out these experiments, according to Menezes and Jaćimović [16], the vials were prepared this way: (i) Vial 1, the large sample (LS): an aliquot of about 5 g of reference material IAEA-SOIL-7 was sealed in a polyethylene vial of 1.3 cm diameter and filled up to 3.6 cm high. The first neutron flux monitor was placed below ampoule, the second inside the ampoule on the top of the sample, and the third outside the vial. Air space between the sample and top of vial was filled with cellulose paper; (ii) Vial 2: six polyethylene vials (9 mm in diameter and 5 mm high) filled with reference material IAEA-SOIL-7 were stacked with neutron flux monitors in sandwich form and inserted into a 5 cm high polyethylene vial; (iii) Vial 3, the small sample (SS): an aliquot of about 200 mg of reference material was sealed in a polyethylene vial (diameter of 9 mm and 5 mm high), stacked together with neutron flux monitors and inserted in a 5-cm-high vial. Air space between the sample and top of ampoule was filled with cellulose paper; (iv) Vial 4: the large sample (LS) was filled in with a soil aliquot, 3.6 cm high and 1.3 cm in diameter. After being prepared, each vial was inserted in the "rabbit". **Figure 1** shows a scheme of the prepared vials.

4.1. Detector efficiency over the volume source

In this step, the influence of distance sample-detector for small and large cylindrical samples based on gamma efficiency was verified. The LS, Vial 1, was measured at 2, 5, 10 and 20 cm sample-detector distances. The gamma efficiency values, ε_p, at sample-detector distances related to several gamma lines (non-coincidence gamma lines) and respective nuclides were given by KayWin software, [39] when calculating the elemental concentration of the large cylindrical sample. The efficiency values were also obtained by ANGLE software [41]. **Table 1** displays examples of the non-coincidence gamma energies for each nuclide determined and the respective ratio of each ε_p to ε_p of the reference distance sample-detector, 20 cm, for KayWin and ANGLE.

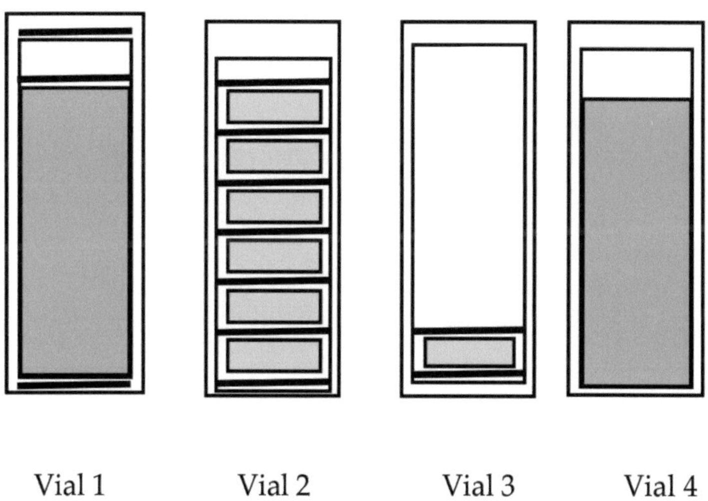

Figure 1. Vials prepared to carry out the experiments: Vial 1, large sample with reference material; Vial 2, with six small samples with reference material; Vial 3, with one small sample filled with reference material and Vial 4, with large sample filled with soil.

Nucl.	Non-coinc. gamma lines (keV)	Large sample ratio (ε_p at specific distance (cm) to ε_p reference distance (cm) sample-detector)							
		KayWin				ANGLE			
		20/20	10/20	5/20	2/20	20/20	10/20	5/20	2/20
Sm-153	103.2	1.000	2.837	6.450	12.749	1.000	2.834	6.461	12.915
Ce-141	145.4	1.000	2.785	6.210	12.079	1.000	2.778	6.214	12.224
Sc-47	159.4	1.000	2.770	6.147	11.902	1.000	2.764	6.150	12.043
Au-199	208.2	1.000	2.734	5.987	11.455	1.000	2.727	5.989	11.589
Ru-97	215.7	1.000	2.730	5.970	11.405	1.000	2.724	5.974	11.546
Pa-233	300.1	1.000	2.699	5.832	11.024	1.000	2.693	5.838	11.163
Pa-233	311.9	1.000	2.697	5.821	10.993	1.000	2.690	5.827	11.132
Cr-51	320.1	1.000	2.695	5.814	10.972	1.000	2.688	5.819	11.111
Nd-147	531	1.000	2.667	5.693	10.637	1.000	2.660	5.695	10.763
As-76	559.2	1.000	2.665	5.682	10.607	1.000	2.657	5.685	10.733
Sb-122	564.2	1.000	2.664	5.680	10.602	1.000	2.657	5.683	10.728
Zn-65	1115.5	1.000	2.635	5.555	10.258	1.000	2.629	5.560	10.386
Fe-59	1291.6	1.000	2.630	5.532	10.194	1.000	2.623	5.538	10.323
K-42	1524.7	1.000	2.624	5.508	10.130	1.000	2.618	5.514	10.258
Sb-124	1691	1.000	2.621	5.495	10.094	1.000	2.615	5.502	10.223

Table 1. Non-coincidence gamma lines: ratio of gamma efficiency (ε_p) determined for a gamma line at a distance to gamma efficiency at reference distance sample-detector for the large sample, reference material IAEA-SOIL-7, Vial 1.

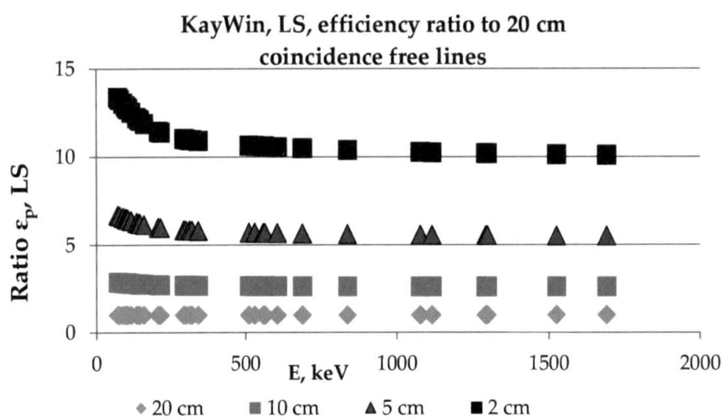

Figure 2. Gamma efficiency normalised to 20 cm obtained by KayWin software for the large cylindrical sample, LS, Vial 1 [16].

Figures 2 and **3** display the gamma efficiency, ε_p, calculated by gamma energy for distance sample-detector normalised to 20 cm for the HPGe detector D4. The figures show a good agreement—after 200 keV—between the ratio's efficiency, ε_p, obtained by KayWin and ANGLE software at 2, 5, 10 and 20 cm and normalised to efficiency calculated at 20 cm.

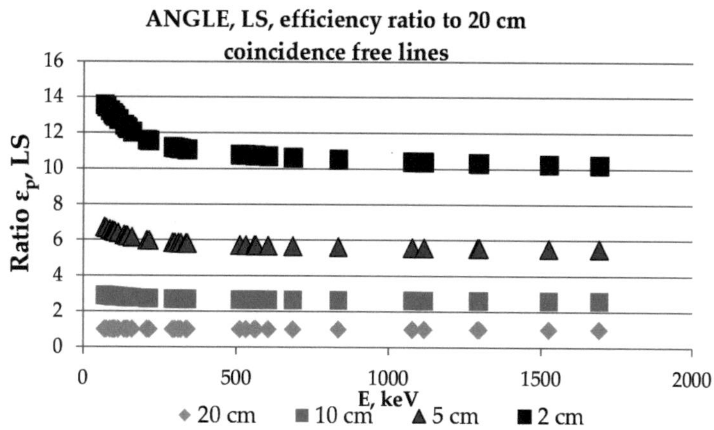

Figure 3. Gamma efficiency normalised to 20 cm obtained by ANGLE software for the large cylindrical sample, LS, Vial 1.

4.2. Neutron self-shielding and the spatial neutron flux distribution factors

4.2.1. Neutron self-shielding

In this methodology development, the KayWin [39] and MATSSF software [45] (specific to calculate the correction factor to thermal and epithermal neutron self-shieldings, G_{th} and G_e, respectively), were applied. A soil sample in cylindrical geometry, large sample, with diameter of 13 mm and 36 mm high was chosen to be studied. The composition, $CaCO_3$ (80%) and SiO_2 (20%), was assumed because these two components are typically major constitutions in soil matrices.

The KayWin calculated G_{th} equal to 0.997, while for the same configuration, the MATSSF obtained G_{th} equal to 0.998. There is a small difference from 1.0 for G_{th}, thermal neutron self-shielding. Due to this, no correction for thermal neutron self-shielding in the geological large sample was necessary. The G_e was considered negligible because of such experimental setup, i.e., $G_e = 1.0$. The same approach has been taken for biological samples studied in this work (see Section 5), where $G_{th} = G_e = 1.0$.

4.2.2. Spatial neutron flux distribution factors

4.2.2.1. Axial neutron flux gradient

To verify the axial and radial distributions of fast and thermal neutron fluxes in the same vial, Vial 4, experiments were carried out and are described in Jaćimović [48], and Menezes and partners [37], based on the experiment performed by Jaćimović and partners [40]. For these experiments, iron wires (99.9% Fe from Mallinckrodt, USA, 0.4 mm in diameter and 5 cm in length) were used.

During irradiation, the reactions were the following:

a. $^{58}Fe(n_t,\gamma)^{59}Fe$, 44.50 day half-life, gamma emissions of 1099.3 keV and 1291.6 keV from ^{59}Fe;

b. $^{54}Fe(n_f,p)^{54}Mn$, 312.2 day half-life, gamma emission of 834.8 keV from ^{54}Mn.

The characterisation of neutron flux gradients (axial and radial) in the irradiation channels in the carousel facility was calculated. The calculation was based on the specific activities of ^{59}Fe based on the mean value of the relative specific activities obtained for both gamma lines of ^{59}Fe (thermal neutrons) and ^{54}Mn (fast neutrons) applying the KayWin software. After decay time, about 2 weeks, the wires were cut into five 1-cm-pieces and submitted to gamma spectrometry on an HPGe detector D4 with 50% relative efficiency. The average specific activity of ^{59}Fe was calculated based on the activity of 1099.3 and 1291.6 keV peaks and for ^{54}Mn, on the peak of 834.8 keV.

One experiment was developed with one big vial filled with different materials, namely density, clay and air. The iron wires were placed in the vial: one in the centre and four near the wall of the polyethylene vial (3 cm in diameter and 5 cm high filled with clay) between the vial wall and the clay. **Figure 4**, position from the bottom (A = 0–1 cm) to top (E = 4–5 cm), related to thermal neutrons, shows the axial distributions for clay-air. The values of ^{59}Fe were normalised to the average value of all the 1 cm Fe pieces. It can be observed that for thermal neutrons, the axial gradient in the geological sample is about 2%/cm. **Figure 5**, also related to thermal neutrons, shows the normalised radial distribution for clay-air. The figure points out that radial gradient for thermal neutrons in clay is about 2%/cm.

4.2.2.2. $F_{c,Au}$-*factor*

It was assumed that the $F_{c,Au}$-factor value is the correspondent to the average height of the sample. The $F_{c,Au}$-factors, calculated by KayWin based on Al-0.1%Au monitors in Vials 1 and 2, were used to verify the trend of axial gradient. Menezes and Jaćimović [16] pointed out that the $F_{c,Au}$-factors have a linear trend.

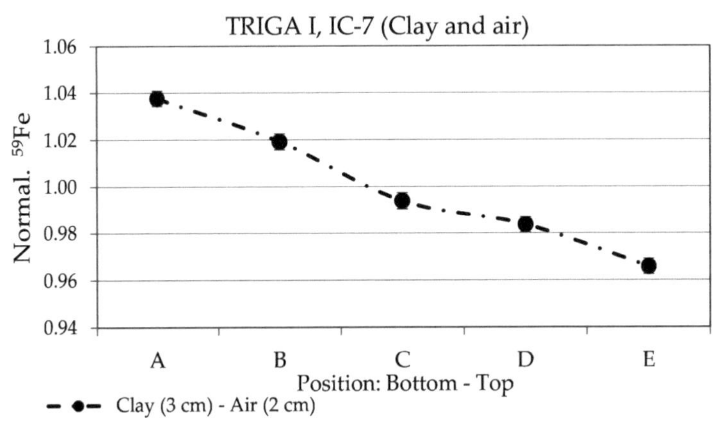

Figure 4. Axial gradient for thermal neutrons in clay-air in the IC-7 irradiation channel of the TRIGA reactor. The values are normalised related to ^{59}Fe and error bars are calculated based on the average statistics counting of 1099.3 and 1291.6 keV from ^{59}Fe, Vial 4 [37].

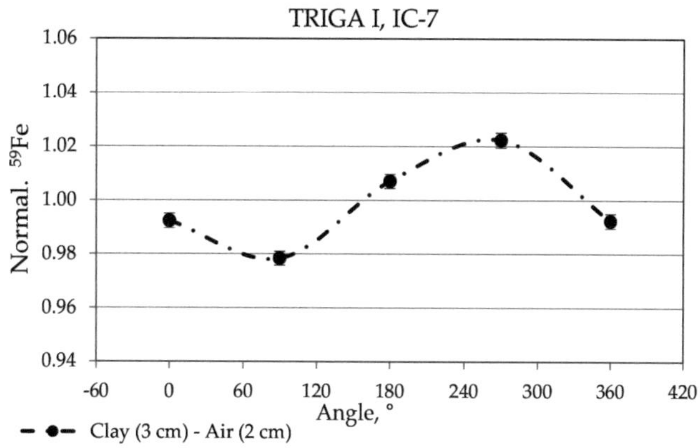

Figure 5. Radial gradient for thermal neutrons in clay-air in the IC-7 irradiation channel of the TRIGA reactor. The values are normalised related to ^{59}Fe and error bars are calculated based on the average statistics counting of 1099.3 and 1291.6 keV from ^{59}Fe, Vial 4 [37].

4.3. Gamma-ray attenuation within the sample during counting

The programme KayWin makes several corrections as already mentioned in the subsection 3.3, Gamma-ray attenuation within the sample during counting, during elemental concentration calculations. As the values obtained for IAEA-SOIL-7 for small and large samples were according to recommended values, according to next subsection 5 (Comparison between small and large samples' elemental concentration results), it was not necessary to develop more tests and apply correction factors.

5. Comparison between small and large samples' elemental concentration results

KayWin software calculated the efficiencies of the large sample accordingly, and the calculated $F_{c,Au}$–factors for the large sample obtained for Al-0.1% Au monitors were described by a linear equation. Then, it was decided to verify the experimental mass fractions for the small (Vial 3) and the large sample (Vial 1). The samples were measured at 2, 5, 10 and 20 cm at detector D4 (50% relative efficiency), and the elemental concentrations were calculated for each distance and for several distances, called RP, a routine procedure of analysis for customers, i.e., suitable distance sample-detector depending on the activity/dead-time. The experimental results and recommended values for reference material IAEA-SOIL-7 calculated by KayWin are shown in Menezes and Jaćimović [16].

Table 2 shows the normalised values of small and large samples to IAEA-SOIL-7 recommended data. It is possible to observe that majority of results are within 95% of confidence interval for assigned values. For the small sample, 88% of the results presented deviations from the recommended values lower than 10%, while for cylindrical samples, the deviations were 74%. For both samples measured at 10 cm, they presented 12% of deviations, and it was the best

El.	Recommended values (mg kg^{-1})	Normalised values on IAEA-SOIL-7 recommended data (date of issue: 2000)									
		Ratio result obtained at specific distance (cm)/recommended value									
		Small sample (~200 mg)					Large sample (~5 g)				
		2	5	10	20	RP	2	5	10	20	RP
As	13.4 ± 0.85	1.11	1.05	1.07	1.04	1.09	1.17	1.15	1.15	1.12	1.15
Ce	61 ± 6.50	1.00	0.97	0.95	0.95	1.00	1.04	0.99	0.98	0.97	1.01
Co	8.9 ± 0.85	1.05	1.03	1.02	1.07	1.05	1.06	1.04	1.04	1.02	1.06
Cr	60 ± 12.5	1.22	1.18	1.16	1.15	1.15	1.30	1.25	1.26	1.22	1.30
Cs	5.4 ± 0.75	1.05	1.05	1.05	1.05	1.05	1.11	1.09	1.09	1.08	1.10
Eu	1.0 ± 0.2	1.02	1.07	1.10	0.85	0.95	1.12	1.17	1.08	1.10	1.12
Hf	5.1 ± 0.35	1.04	1.02	1.01	1.01	1.04	1.05	1.04	1.04	1.01	1.03
La	28 ± 1	1.06	1.01	1.02	1.01	1.04	1.07	1.05	1.04	1.03	1.04
Nd	30 ± 6	0.87	0.94	0.92	0.95	0.88	0.98	1.10	1.02	1.09	1.08
Rb	51 ± 4.5	1.04	0.99	0.97	0.97	1.04	1.09	1.03	1.03	1.04	1.04
Sb	1.7 ± 0.2	1.08	1.06	1.08	1.07	1.08	1.07	1.03	1.02	1.03	1.03
Sc	8.3 ± 1.05	1.12	1.10	1.08	1.08	1.12	1.13	1.11	1.10	1.09	1.11
Sm	5.1 ± 0.35	0.96	0.98	0.93	0.92	0.96	1.03	1.01	0.98	0.96	1.01
Ta	0.8 ± 0.2	0.91	0.94	0.93	0.93	0.91	0.95	0.98	0.98	0.97	0.95
Tb	0.6 ± 0.2	1.10	1.13	1.10	1.10	1.10	1.22	1.19	1.14	1.15	1.21
Th	8.2 ± 1.1	1.02	1.01	0.98	0.97	1.02	1.08	1.06	1.05	1.03	1.08
U	2.6 ± 0.55	0.90	0.85	0.87	0.85	0.85	0.99	0.90	0.91	0.89	0.90
Yb	2.4 ± 0.35	1.00	0.98	0.95	0.94	1.00	1.06	0.99	0.99	0.98	0.99
Zr	185 ± 10.5	1.12	1.11	1.13	1.12	1.13	1.29	1.14	0.94	1.06	1.30

El., element; RP, calculations carried out according to the usual procedure to the customers.

Table 2. Normalised values of small (Vial 3) and large samples (Vial 1) obtained at specific distance to IAEA-SOIL-7 on recommended data [16].

performance. Kennedy and St-Pierre [49] in their work observed a similar conclusion, where 10 cm distance sample-detector shows the best performance. The authors used an HPGe detector with 29% relative efficiency. Same conclusion can be explained due to the same method used for the absolute calibration procedure of an HPGe detector. Nevertheless, results presented in **Table 2** for IAEA-SOIL-7 show relatively small differences between measurement distance sample-detector. So, it is possible to conclude that the impact from different measurement positions (or ratio between measurement and reference position) contributed relatively small systematic error to the final result.

The analytical performance of the experiments and the agreement of element contents with recommended values, the assigned values for IAEA-SOIL-7, i.e., the data given in the Certificate of analysis in 2000 [47] were evaluated by the statistical test called E_n-*score* [50]. This score takes into account the expanded uncertainty of both values with a coverage factor $k = 2$ (95% confidence interval). At CDTN, the uncertainty of the k_0-NAA established is considered as 3.5% with a coverage factor $k = 1$. E_n-*score* is calculated as follows:

$$E_n = \frac{X_{lab} - X_{ref}}{\sqrt{U_{lab}^2 + U_{ref}^2}} \tag{4}$$

where X_{lab} and X_{ref} are laboratory and reference (assigned value) values, respectively; U_{lab} and U_{ref} are expanded uncertainties with a coverage factor $k = 2$ of laboratory and reference values, respectively. The criterion $|E_n| \leq 1$ was applied to compare the results of the two geometries with reference data, meaning that the evaluation of the performance of the method was satisfactory, producing values with a level of confidence of about 95% and if $|E_n| > 1$, the performance was unsatisfactory. **Table 3** displays the results for this test.

Element	**Recommended values (mg kg^{-1})**	**Experimental values**			
		Small sample		**Large sample**	
		(mg kg^{-1})	$\|E_n\|$	**(mg kg^{-1})**	$\|E_n\|$
As	13.4 ± 0.85	14.2 ± 0.5	0.59	14.2 ± 0.5	0.56
Ce	61 ± 6.5	60.8 ± 2.1	0.03	59.9 ± 2.1	0.13
Co	8.9 ± 0.85	9.33 ± 0.33	0.40	9.17 ± 0.32	0.25
Cr	60 ± 12.5	68.9 ± 4.0	0.60	63.6 ± 2.6	0.27
Cs	5.4 ± 0.75	5.69 ± 0.20	0.35	5.83 ± 0.21	0.50
Eu	1.0 ± 0.2	0.95 ± 0.03	0.24	1.12 ± 0.16	0.33
Hf	5.1 ± 0.35	5.33 ± 0.20	0.43	5.34 ± 0.19	0.46
La	28 ± 1	29 ± 1	0.44	29 ± 1	0.42
Nd	30 ± 6	26.3 ± 1.2	0.57	27.9 ± 1.3	0.32
Rb	51 ± 4.5	53.0 ± 2.4	0.31	51.8 ± 2.5	0.12
Sb	1.7 ± 0.2	1.83 ± 0.07	0.54	1.73 ± 0.06	0.14
Sc	8.3 ± 1.05	9.29 ± 0.33	0.80	9.09 ± 0.32	0.65
Sm	5.1 ± 0.35	4.87 ± 0.19	0.44	5.22 ± 0.18	0.24
Ta	0.8 ± 0.2	0.73 ± 0.03	0.33	0.78 ± 0.03	0.11
Tb	0.6 ± 0.2	0.66 ± 0.02	0.30	0.69 ± 0.02	0.45
Th	8.2 ± 1.1	8.35 ± 0.29	0.12	8.61 ± 0.30	0.32
U	2.6 ± 0.55	2.21 ± 0.13	0.64	2.55 ± 0.10	0.09
Yb	2.4 ± 0.35	2.39 ± 0.09	0.03	2.37 ± 0.09	0.08

Uncertainties of recommended values are given at a confidence interval of 95% ($k = 2$), while experimental values are given as combined standard uncertainty.

Table 3. Experimental results and recommended values for IAEA-SOIL-7 analysed as small (Vial 3) and large samples (Vial 1) [37].

6. Some applications

In order to confirm the applicability and its ability to produce good results, several reference materials were analysed as small and large samples. **Tables 4–6** show the results of certified

BCR-320R, ***channel sediment***					
Element	**Certified values, k = 2**	**Experimental results, k = 1**			
		Small sample, n = 3, (~0.150 g)		**Large samples, n = 3, (~1 g)**	
	(mg kg^{-1})	**(mg kg^{-1})**	$\lvert E_n \rvert$	**(mg kg^{-1})**	$\lvert E_n \rvert$
As	21.7 ± 2.0	21.2 ± 0.7	0.19	21.5 ± 0.9	0.08
Co	9.7 ± 0.6	9.3 ± 0.3	0.48	9.2 ± 0.3	0.54
Cr	59 ± 4	58 ± 2	0.19	57 ± 3	0.39
Fe	25,700 ± 1300	24,180 ± 851	0.71	24,100 ± 867	0.74
Sc	5.2 ± 0.4	5.0 ± 0.2	0.41	5.0 ± 0.2	0.37
Th	5.3 ± 0.4	5.0 ± 0.2	0.51	4.9 ± 0.2	0.68
U	1.56 ± 0.20	1.5 ± 0.1	0.29	1.4 ± 0.1	0.49
Zn	319 ± 20	304 ± 11	0.48	310 ± 11	0.28

n, number of replicates.

Table 4. Mass fractions obtained for certified reference material BCR-320R, *Channel Sediment*, analysed as small and large samples [51].

Element	**NIST SRM 1572,** ***citrus leaves***				
	Certified values, k = 2	**Experimental results, k = 1**			
		Small sample (0.2124 g)		**Large sample (2.4758 g)**	
	(mg kg^{-1})	**(mg kg^{-1})**	$\lvert E_n \rvert$	**(mg kg^{-1})**	$\lvert E_n \rvert$
As	3.1 ± 0.3	3.5 ± 0.1	0.91	2.8 ± 0.1	0.78
Ba	21 ± 3	21 ± 2	0.09	18 ± 1	0.97
Ca	31,500 ± 1000	34,780 ± 1554	1.00	29,150 ± 1098	0.97
Cr	0.8 ± 0.2	0.8 ± 0.1	0.12	0.81 ± 0.05	0.05
Fe	90 ± 10	108 ± 8	0.95	98 ± 4	0.58
Hg	0.08 ± 0.02	< 0.2	—	0.09 ± 0.01	0.42
K	18,200 ± 600	20,570 ± 1200	0.96	17,360 ± 684	0.56
Na	160 ± 20	185 ± 7	1.00	165 ± 6	0.22
Rb	4.84 ± 0.06	5.3 ± 0.5	0.54	4.5 ± 0.2	0.90
Sr	100 ± 2	111 ± 8	0.73	94 ± 4	0.61

Table 5. Mass fractions obtained for certified reference material NIST SRM 1572, *Citrus Leaves*, analysed as small and large samples [52].

Element	NIES no 7, *tea leaves*				
	Certified values, $k = 2$	Experimental results, $k = 1$			
		Small sample (0.3024 g)		Large sample (3.5517 g)	
	(mg kg^{-1})	(mg kg^{-1})	$\|E_n\|$	(mg kg^{-1})	$\|E_n\|$
Br	2.5 ± 0.1	2.8 ± 0.2	0.77	2.4 ± 0.1	0.56
Ca	3200 ± 120	3607 ± 266	0.75	2794 ± 200	0.97
Fe	98 ± 7	114 ± 9	0.84	98 ± 5	0.01
K	18,600 ± 700	20,240 ± 800	0.94	18,200 ± 687	0.26
La	0.068 ± 0.002	0.080 ± 0.006	1.00	0.064 ± 0.003	0.58
Na	15.5 ± 1.5	18 ± 1	0.97	18 ± 1	1.00
Rb	6.59 ± 0.01	6.7 ± 0.3	0.14	5.8 ± 0.5	0.93

Table 6. Mass fractions obtained for reference material NIES no 7, *tea leaves,* analysed as small and large samples [52].

El.	Oatmeal powder			Chia powder		
	Small sample (0.2040 g)	Large sample (0.6085 g)	Large sample (2.5396 g)	Small sample (0.2415 g)	Large sample (0.5703 g)	Large sample (3.4228 g)
	(mg kg^{-1})	(mg kg^{-1})	(mg kg^{-1})	(mg kg^{-1})	(mg kg^{-1})	(mg kg^{-1})
Ba	<10	8 ± 1	8 ± 1	45 ± 2	45 ± 2	45 ± 2
Br	1.88 ± 0.07	1.81 ± 0.06	1.79 ± 0.06	8.7 ± 0.3	8.8 ± 0.3	8.0 ± 0.3
Ca	<690	643 ± 127	553 ± 87	7081 ± 504	7281 ± 462	6918 ± 572
Co	0.024 ± 0.007	0.029 ± 0.002	0.030 ± 0.003	0.35 ± 0.01	0.35 ± 0.02	0.37 ± 0.02
Cs	0.087 ± 0.005	0.092 ± 0.004	0.086 ± 0.004	0.044 ± 0.004	0.044 ± 0.003	0.044 ± 0.003
Fe	56 ± 7	60 ± 3	56 ± 3	118 ± 6	97 ± 5	109 ± 5
K	4134 ± 146	4175 ± 148	3972 ± 140	8531 ± 300	8618 ± 304	8213 ± 291
La	<0.01	<0.01	<0.01	0.069 ± 0.006	0.062 ± 0.004	0.073 ± 0.003
Mo	0.5 ± 0.1	<0.6	0.6 ± 0.1	<0.4	<0.4	<0.4
Na	10 ± 1	10.6 ± 0.4	10.9 ± 0.4	7.3 ± 0.2	7.2 ± 0.3	8.8 ± 0.4
Rb	9.0 ± 0.4	8.7 ± 0.4	8.3 ± 0.3	11.8 ± 0.5	11.8 ± 0.5	11.9 ± 0.5
Sc*	2.9 ± 0.4	1.8 ± 0.3	1.6 ± 0.3	<0.02	<0.02	<0.02
Sm	<0.004	<0.004	<0.004	<0.008	0.008 ± 0.002	0.010 ± 0.001
Sr	<10	<10	<10	46 ± 3	47 ± 3	45 ± 2
Ta	<0.01	<0.005	0.018 ± 0.004	<0.03	<0.03	<0.03
Zn	<32	<32	<32	61 ± 2	61 ± 2	51 ± 2

El., element; experimental values are given as combined standard uncertainty.
*µg kg^{-1}.

Table 7. Mass fractions obtained for oatmeal powder and chia powder analysed in three sizes of samples: one small and two large samples [53].

Element	Soil sample				
	Small sample (0.1133 g) (mg kg^{-1})	Large sample (0.6313 g) (mg kg^{-1})	Element	Small sample (0.1133 g) (mg kg^{-1})	Large sample (0.6313 g) (mg kg^{-1})
Fe	23,070 ± 952	22,100 ± 1235	Ta	2.31 ± 0.09	2.26 ± 0.12
Ga	37.2 ± 1.3	33.9 ± 1.2	Tb	0.503 ± 0.032	0.515 ± 0.019
Hf	32.8 ± 1.5	30.4 ± 2.0	Th	31.0 ± 1.3	29.9 ± 1.5
K	347 ± 30	367 ± 25	U	3.71 ± 0.14	3.49 ± 0.15
La	70.3 ± 2.7	70.0 ± 2.5	W	13.4 ± 0.5	13.7 ± 0.5
Sb	0.471 ± 0.046	0.430 ± 0.038	Yb	5.61 ± 0.36	5.21 ± 0.15
Sc	10.4 ± 0.4	9.7 ± 0.5	Zn	15.2 ± 1.9	16.0 ± 2.1

Experimental values are given as combined standard uncertainty.

Table 8. Mass fractions obtained for soil, analysed as small and large samples.

reference materials, BCR-320R, *Channel Sediment*, NIST SRM 1572, *Citrus Leaves*, and reference material, NIES no 7, *Tea Leaves*, respectively. **Tables 7** and **8** show examples of samples analysed at the Laboratory for Neutron Activation Analysis, as small and large samples.

7. Conclusions

The Laboratory for Neutron Activation Analysis at CDTN, Brazil, needs to be prepared to meet customer analysis requirements. In order to answer this requirement, it was necessary to establish a methodology to analyse large samples, from 0.5 to 5 g sample, applying the k_0-standardised method keeping the current infrastructure for irradiation and gamma measurements. Another objective was to overcome technical limitations while analysing samples smaller than 0.5 g, as, for instance, inhomogeneity of the sample and lower thermal neutron flux of the reactor.

To carry out the establishment, it was necessary to calculate several relevant parameters that would point out the feasibility to analyse large samples: detector efficiency over the volume source, neutron self-shielding during neutron irradiation and gamma-ray attenuation within the sample during counting.

The results [16, 36, 37] pointed out that all parameters may be negligible for the large sample. At least, within the uncertainty range, once the elemental concentration results are in good agreement with the recommended values of the reference material IAEA-SOIL-7. All results obtained during the development using the reference material were evaluated by the statistical test, E_n-*score*, which pointed out that the k_0-method applied using KayWin software presented an overall satisfactory performance. Few exceptions are already being investigated using other reference materials. It means that the KayWin software proved to be a robust programme to calculate the elemental concentration of large samples producing reliable results.

It is important to note that once the methodology is established, the following benefits may be obtained: reaching lower limits of detection; enabling compliance, for example, in environmental legislation in determining the concentration of metals in soil; analyses of more representative aliquots, especially in the case of non-homogeneous samples, as industrial waste; optimization of the cost and time of analysis, because instead of analysing several small samples they could be replaced by a large sample; analyses of whole parts when removing the aliquots are not possible. This study is still going on in order to confirm the benefits that have been already mentioned.

This investigation has confirmed that any other laboratory applying k_0-instrumental neutron activation analysis (k_0-INAA) is able to establish this methodology without having to modify its facilities, since the neutron self-shielding, gamma attenuation and detector efficiency over the volume sample are established.

Acknowledgements

This work was partially supported by the International Atomic Energy Agency under grant BRA-14798, by the Brazilian Foundation for Research Support of Minas Gerais, FAPEMIG, under grant APQ-01259-09, and by financial support from the Slovenian Research Agency (ARRS) through programme P1-0143. Thanks to Brazilian National Council for Scientific and Technological Development, CNPq, and to Coordination for the Improvement of Higher Education Personnel, CAPES. Thanks to the TRIGA MARK I IPR-R1 reactor staff for making the use of the reactor for the experiments possible. Last but not least thanks to Dr Tibor G. Kocsor (*Journal of Radioanalytical and Nuclear Chemistry*) for allowing the use of tables and figures previously published.

Author details

Maria Ângela de B.C. Menezes[1]* and Radojko Jaćimović[2]

*Address all correspondence to: menezes@cdtn.br

1 Nuclear Technology Development Centre, Brazilian Commission for Nuclear Energy, CDTN/CNEN, Belo Horizonte, Minas Gerais, Brazil

2 Jožef Stefan Institute, Ljubljana, Slovenia

References

[1] Hamidatou L, Slamene H, Akhal T, Zouranen B. Concepts, Instrumentation and techniques of neutron activation analysis. In: Kharfi F, editor. Imaging and Radioanalytical Techniques in Interdisciplinary Research. IntechOpen. DOI: 10.5772/53686. Available

from: https://www.intechopen.com/books/imaging-and-radioanalytical-techniques-in-interdisciplinary-research-fundamentals-and-cutting-edge-applications/concepts-instrumentation-and-techniques-of-neutron-activation-analysis

[2] De Soete D, Gijbels R, Hoste J. Neutron activation analysis. A Series of Monographs on Analytical Chemistry and Its Applications Chemical Analysis. Vol. 34. Wiley-Insterscience; 1972. p. 836

[3] Bode P. Opportunities for innovation in neutron activation analysis. Journal of Radioanalytical and Nuclear Chemistry. 2012;**291**:275-280

[4] Greenberg RR, Bode P, Fernandes EAN. Neutron activation analysis: A primary method of measurement. Spectrochimica Acta, Part B: Atomic Spectroscopy. 2011;**66**:193-241

[5] De Corte F, Simonits A, De Wispelaere A, Hoste J. Accuracy and applicability of the k_0-standardization method. Journal of Radioanalytical and Nuclear Chemistry. 1987;**113**:145-161

[6] Jaćimović R, Smodiš B, Bučar T, Stegnar P. k_0-NAA quality assessment by analysis of different certified reference materials using the KAYZERO/SOLCOI software. Journal of Radioanalytical and Nuclear Chemistry. 2003;**257**:659-663

[7] Leal AS, Menezes MABC, Jaćimović R, Sepe FP, Gomes TCB. Investigation of the quality of the drug Enalapril commercialized by pharmacies of manipulation in Belo Horizonte, Brazil. Journal of Radioanalytical and Nuclear Chemistry. 2014;**300**:645-651

[8] Menezes MABC, Sabino CVS, Franco MB, Kastner GF, Rossi Montoya EH. k_0-Instrumental Neutron Activation analysis establishment at CDTN, Brazil: A successful story. Journal of Radioanalytical and Nuclear Chemistry. 2003;**257**:627-632

[9] Menezes MABC, Jaćimović R. Optimised k_0-instrumental neutron activation method using the TRIGA MARK I IPR-R1 reactor at CDTN/CNEN, Belo Horizonte, Brazil. Nuclear Instruments and Methods in Physics Research A. 2006;**564**:707-715

[10] Menezes MABC, Maia ECP, Albinati CCB, Sabino CVS, Batista JR. How suitable are scalp hair and toenail as biomonitors? Journal of Radioanalytical and Nuclear Chemistry. 2004;**259**:81-86

[11] Menezes MABC, Maia ECP, Albinati CCB. Instrumental neutron activation analysis as an analytical tool on supporting the establishment of guidelines and data basis for worker's health awareness program. Revista Brasileira de Pesquisa e Desenvolvimento. 2002;**4**:1110-1117

[12] Menezes MABC, Palmieri HEL, Leonel LV, Nalini HA Jr, Jaćimović R. Iron Quadrangle, Brazil: Elemental concentration determined by k_0-instrumental neutron activation analysis Part I: soil samples. Journal of Radioanalytical and Nuclear Chemistry. 2006;**270**:111-116

[13] Menezes MABC, Palmieri HEL, Leonel LV, Nalini HA Jr, Jaćimović R. Elemental concentration determined by k_0-instrumental neutron activation analysis Part II: kale samples. Journal of Radioanalytical and Nuclear Chemistry. 2006;**270**:117-121

[14] Menezes MABC, Pelaes ACO, Salles PMB, Silva WF, Moura RR, Moura IFS, Jaćimović R. Neutron activation technique: A reliable tool to determine the mineral composition in agro-industrial products. Radiation and Applications. 2017;**2**:124-128

[15] Menezes MABC, Sabino CVS, Franco MB, Maia ECP, Albinati CCB. Assessment of Workers' Contamination caused by air pollution exposure in industry using biomonitors. Journal of Atmospheric Chemistry. 2004;**49**:403-414

[16] Menezes MABC, Jaćimović R. Implementation of a methodology to analyse cylindrical 5-g sample by neutron activation technique, k_0 method, at CDTN/CNEN, Belo Horizonte, Brazil. Journal of Radioanalytical and Nuclear Chemistry. 2014;**300**:523-531

[17] Salles PMB, Menezes MABC, Sathler MM, Moura RR, Campos TPR. Inorganic elements in sugar samples consumed in several countries. Journal of Radioanalytical and Nuclear Chemistry. 2015;**306**:1-9

[18] Salles PMB, Menezes MABC, Sathler MM, Moura RR, Campos TPR. Elemental composition of dietary supplements most consumed in Belo Horizonte, Brazil, analysed by k_0-INAA. Journal of Radioanalytical and Nuclear Chemistry. 2017;**308**:1-5

[19] Silveira JN, Menezes MABC, Lara PCP, Beinner MA, Silva JBB. Determination of Al, Ca, Cl, Cr, K, Mg, Sb and Ti in industrialized and formulated antihypertensive drugs using Neutron Activation Analysis. Journal of Pharmacy and Biological Sciences. 2015;**10**:40-45

[20] Viana CO, Menezes MABC, Maia ECP. Epiphytic lichens on air biomonitoring in Belo Horizonte City, Brazil: A preliminary assessment. International Journal of Environment and Health. 2011;**5**:324-336

[21] De Corte F. The k_0–Standardisation Method: A Move to the Optimisation of Neutron Activation Analysis [thesis]. Ryksuniversiteit Ghent, Faculteit Van de Wetenschappen; 1987

[22] De Corte F, Simonits A. Recommended nuclear data for use in the k_0 standardization of neutron activation analysis. Atomic Data and Nuclear Data Tables. 2003;**85**:47-67

[23] Jaćimović R, De Corte F, Kennedy G, Vermaercke P, Revay Z. The 2012 recommended k_0 database. Journal of Radioanalytical and Nuclear Chemistry. 2014;**300**:589-592

[24] k_0-database 2015 [Internet]. 2015. Available from: http://www.kayzero.com/k0naa/k0naaorg/Nuclear_Data_SC/Entries/2016/1/11_New_k0-data_Library_2015.html [Accessed: 25-10-2018]

[25] Bode P, Overwater RMW, De Goeij JJM. Large-sample neutron activation analysis: Present status and prospects. Journal of Radioanalytical and Nuclear Chemistry. 1997;**216**:5-11

[26] Gwozdz R. Instrumental neutron activation analysis of samples with volumes from 2 to 350 ml. Journal of Radioanalytical and Nuclear Chemistry. 2007;**271**:751-759

[27] Tzika F, Stamatelatos IE, Kalef-Ezra J. Neutron activation analysis of large samples: The influence of inhomogeneity. Journal of Radioanalytical and Nuclear Chemistry. 2007;**271**:233-240

[28] Blaauw M, Lakmaker O, Van Aller P. The accuracy of instrumental neutron activation analysis of kilogram-size inhomogeneous samples. Analytical Chemistry. 1997;**69**:2247-2250

[29] Shakir NS, Jervis RE. Correction factors required for quantitative large volume INAA. Journal of Radioanalytical and Nuclear Chemistry. 2001;**248**:61-68

[30] Lin X, Henkelmann R. Instrumental neutron activation analysis of large samples: A pilot experiment. Journal of Radioanalytical and Nuclear Chemistry. 2002;**251**:197-204

[31] Overwater RMW. The Physics of Big Sample Instrumental Neutron Activation Analysis [thesis]. Delft: Delft University of technology; 1994

[32] Stamatelatos IE, Tzika F. Large sample neutron activation analysis: A challenge in cultural heritage studies. Annali di Chimica. 2007;**97**:505-512

[33] Lindstrom RM, Fleming RF. Neutron self-shielding factors for simple geometries, Revisited. Chemist-Analyst. 2008;**53**:855-859

[34] Musa Y, Ahmed YA, Yamusa YA, Ewa IOB. Determination of radial and axial neutron flux distribution in irradiation channel of NIRR-1 using foil activation technique. Annals of Nuclear Energy. 2012;**50**:50-55

[35] Tzika F, Stamatelatos IE, Kalef-Ezra J, Bode P. Large sample neutron activation analysis: Correction for neutron and gamma attenuation. Nukleonika. 2004;**49**:115-121

[36] Menezes MABC, Jaćimović R, Ribeiro L. Contribution of analytical nuclear techniques in the reconstruction of the Brazilian prehistory analysing archaeological ceramics of Tupiguarani tradition. In: International Atomic Energy Agency, editor. Advances in Neutron Activation Analysis of Large Objects with Emphasis on Archaeological Examples. Research Project. IAEA-TECDOC-1838; 2018. pp. 4-21

[37] Menezes MABC, Jaćimović R, Pereira C. Spatial distribution of neutron flux in geological larger sample analysis at CDTN/CNEN, Brazil. Journal of Radioanalytical and Nuclear Chemistry. 2015;**306**:611-616

[38] HyperLab Gamma Spectroscopy Software, HyperLabs, Software [Internet]. 2009. Available from: http://hlabsoft.com/ [Accessed: 09-06-2013]

[39] Van Sluijs R. Kayzero for Windows User's Manual, for reactor neutron activation analysis (NAA) using the k_0 standardisation method, V. 2. Software developed by DSM Research, Geleen (NL) for NAA based on the k_0 standardization method developed at the INW-RUG, Gent (B) and the AEKI, Budapest (H). k_0-Ware, Heerlen, The Netherlands, 2011. Available: http://www.kayzero.com/KfW%20Manual%20V1.pdf [Accessed: 2013-06-12]

[40] Jaćimović R, Stibilj V, Benedik L, Smodiš B. Characterization of the neutron flux gradients in typical irradiation channels of a TRIGA Mark II reactor. Journal of Radioanalytical and Nuclear Chemistry. 2003;**257**:545-549

[41] Jovanović S, Dlabač A, Mihaljević N, Vukotić P. ANGLE: A PC-code for semiconductor detector efficiency calculations. Journal of Radioanalytical and Nuclear Chemistry. 1997;**218**:13-20

[42] Matsushita R, Koyama M, Yamada S, Kobayashi M, Moriyama H. Neutron flux gradients and spectrum changes in the irradiation capsule for reactor neutron activation analysis. Journal of Radioanalytical and Nuclear Chemistry. 1997;**216**:95-99

[43] Arbocco FF, Vermaercke P, Sneyers L, Strijckmans K. Experimental validation of some thermal neutron self-shielding calculation methods for cylindrical samples in INAA. Journal of Radioanalytical and Nuclear Chemistry. 2011;**291**:529-534. DOI: 10.1007/s10967-011-1211-y

[44] Chilian C, St-Pierre J, Kennedy G. Dependence of thermal and epithermal neutron self-shielding on sample size and irradiation site. Nuclear Instruments and Methods in Physics Research Section A: Accelerators, Spectrometers, Detectors and Associated Equipment. 2006;**564**:629-635

[45] Trkov A, Žerovnik G, Snoj L, Ravnik M. On the self-shielding factors in neutron activation analysis. Nuclear Instruments and Methods in Physics Research A. 2009;**610**:553-565

[46] Moens L, De Donder J, Xi-lei L, De Corte F, De Wispelaere A, Simonits A, Hoste J. Calculation of the absolute peak efficiency of gamma-ray detectors for different counting geometries. Nuclear Instruments and Methods in Physics Research. 1981;**187**:451-472

[47] Analytical Quality Control Services (AQCS) Agency's Laboratories. International Atomic Energy Agency Certified reference material IAEA-SOIL-7. IAEA, Vienna, 2000

[48] Jaćimović R. Activities Developed During the Period of the Fellowship BEV, Process CNPq 170058/2011–5 and Institutional Process 551042/2011-7 [report]. Belo Horizonte: Nuclear Technology Development Centre (Centro de Desenvolvimento da Tecnologia Nuclear, CDTN); 2011

[49] Kennedy G, St-Pierre J. Is the k_0 method accurate for elements with high Q_0 values? Journal of Radioanalytical and Nuclear Chemistry. 2003;**257**:475-480

[50] ISO 13528. Statistical Methods for Use in Proficiency Testing by Interlaboratory Comparisons. 2nd ed. Geneva (CH): International Organization for Standardization; 2015

[51] Pelaes AC. Avaliação da incerteza de medição da composição de amostras geológicas analisadas por meio do método k_0 de ativação neutrônica (Evaluation of the uncertainty of the measurement of geological samples analysed by neutron activation analysis, k_0-method). [dissertation], Belo Horizonte: Postgraduate Program in Radiation, Minerals and Materials Science and Technology, CDTN; 2018

[52] Sathler MM, Salles PMB, Oliveira HS, Menezes MABC. Amostras grandes analisadas por ativação neutrônica, método k_0 (Large samples analysed by neutron activation analysis, k_0 method). In: Proceedings of the Quarta Semana de Engenharia Nuclear e Ciências das Radiações (Forth Week of Nuclear Engineering and Sciences of Radiation) (IV SENCIR 2018); Belo Horizonte, 6–8 November 2018. Belo Horizonte: PPCTN, UFMG; 2018. pp. 107-112

[53] Sathler MM. Avaliação da composição elementar de alimentos integrais e refinados por meio do método k_0 de ativação neutrônica aplicado a amostras grandes (Evaluation of the elemental composition of integral and refined foods by k_0-neutron activation method applied to large samples). [dissertation], Belo Horizonte: Postgraduate Program in Radiation, Minerals and Materials Science and Technology, CDTN; 2018